Conceptual Realism and Historicity: Brandom versus Hegel

Value Inquiry Book Series

Founding Editor

Robert Ginsberg

Editor-in-Chief

J.D. Mininger

VOLUME 403

Cognitive Science

Edited by

Francesc Forn i Argimon, *University of Catalonia*

The titles published in this series are listed at *brill.com/vibs* and *brill.com/cosc*

Conceptual Realism and Historicity: Brandom versus Hegel

By

Norman Schultz

BRILL

LEIDEN | BOSTON

Cover illustration: Image created using Microsoft AI Image Generator by Norman Schultz (2024).

Library of Congress Cataloging-in-Publication Data

Names: Schultz, Norman, author.
Title: Conceptual realism and historicity : Brandom versus Hegel / by Norman Schultz.
Description: Leiden ; Boston : Brill, [2025] | Series: Value inquiry book series, 0929–8436 ; volume 403 | Includes bibliographical references and index.
Identifiers: LCCN 2024046875 | ISBN 9789004714793 (hardback) | ISBN 9789004714809 (e-book)
Subjects: LCSH: Hegel, Georg Wilhelm Friedrich, 1770-1831. Phänomenologie des Geistes. | Brandom, Robert. Spirit of trust. | Realism. | Conceptualism. | Historicism. | Analysis (Philosophy) | Philosophy, Modern.
Classification: LCC B2929 .S37 2025 | DDC 193–dc23/eng/20241114
LC record available at https://lccn.loc.gov/2024046875

Typeface for the Latin, Greek, and Cyrillic scripts: "Brill". See and download: brill.com/brill-typeface.

ISSN 0929-8436
ISBN 978-90-04-71479-3 (hardback)
ISBN 978-90-04-71480-9 (e-book)
DOI 10.1163/9789004714809

This book is printed on acid-free paper and produced in a sustainable manner.

Contents

Acknowledgments

I am deeply grateful to all those who have supported me in completing this book over many years. It is a revised, expanded, and restructured version of my dissertation at Duquesne University. Therefore, I would like to extend my heartfelt thanks to my former dissertation director, James Swindal, whose initial philosophical guidance and enduring encouragement were instrumental in bringing this project to fruition. I also express my appreciation to Tom Rockmore for his invaluable and persistent support, as well as his critical advice. My gratitude extends to Charles Don Keyes and Dan Selcer for generously offering their time and providing insightful assessments of the ideas presented in this work. Special thanks are due to RJ Wighton for his meticulous editing and engaging discussions about the book.

Moreover, I am profoundly grateful for my mother's unwavering support, as well as for the contributions of numerous friends who have shaped the ideas in this book, even though I cannot mention each of them by name.

CHAPTER 1

Introduction—The Divide and Brandom's Interpretation of Hegel

Over the past 70 years, the landscape of philosophy has been shaped by the ongoing division between analytic and Continental philosophy.[1] Rorty, who traversed between the two philosophical camps, humorously depicted the conflict early on as "explicit […] when analytic philosophers mutter about the irrationalism that rages in literature departments and when Continental philosophers become shrill about the lack of human significance in the works of the analyst" (Rorty 1982a, 199).[2] In less ironic terms, the divide can be roughly characterized by analytic philosophy's original premise of a non-metaphysical and non-dogmatic analysis of language,[3] prioritizing "clarity and validity" of arguments, whereas Continental philosophers explore a range of interconnected themes, often related to individual experience (see Strassfeld 2020, 833).[4]

The analytic and Continental traditions undoubtedly encompass a multitude of diverse philosophies that resist such easy categorization. In the 1980s,

1 The historical divide between the analytical and Continental traditions can be traced back to earlier years. For instance, it is widely acknowledged that Husserl was aware of both philosophical camps and maintained contact with philosophers from the Vienna Circle (see Ferraris, Maurizio. "Introduction." *Bridging the Analytical Continental Divide: A Companion to Contemporary Western Philosophy*. Edited by Tiziana Andina. Leiden: Brill, 2014, 6). The authors of *Beyond the Analytic-Continental Divide: Pluralist Philosophy in the Twenty-First Century* (Bell, Jeffrey A., Andrew Cutrofello, and Paul M. Livingston. New York: Routledge, 2016) locate the division no earlier than in the 1960s, and Hilary Putnam notes that there was no divide at the end of the 1940s (see Putnam, Hilary. "A Half Century of Philosophy, Viewed from Within." *American Academic Culture in Transformation: Fifty Years, Four Disciplines*. Edited by Thomas Bender and Carl Schorske. Princeton: Princeton University Press, 1998, 193).

2 Rorty, Richard. "Philosophy in America Today." *The American Scholar* 51 (2), 1982: 183–220.

3 Additional characteristics are "varying but overlapping commitments to the linguistic turn, the rejection of metaphysics, the claim that philosophy is continuous with science, a reductive approach to analysis, the employment of formal logic, a focus on argument and a concern for clarity" (Chase, James, and Jack Reynolds. *Analytic Versus Continental: Arguments on the Methods and Value of Philosophy*. New York: Routledge, 2014, 7).

4 The approaches of phenomenology and existentialism stand in the center of Continental philosophy, while on its borders post-structuralism, Hegelianism, and the Frankfurt school reside (see Strassfeld, Jonathan. "American Divide: The Making of 'Continental' Philosophy." *Modern Intellectual History* 17 (3), 2020: 833–866, 833. https://doi.org/10.1017/S1479244318000513).

 DOI:10.1163/9789004714809_002

Rorty's exposure to the wide-ranging perspectives and approaches led him to even question the existence of whether there is any "consensus about the problems and methods of philosophy" at all (Rorty 1982a, 187). Since then, however, there has been an interpretive shift from perceiving philosophy as a fragmented or divided discipline to recognizing its pluralistic nature, capable of accommodating various approaches. As a result, a growing number of philosophers argue that the "distinction between analytic and continental philosophy [...] is increasingly obsolete" (Marshall 2013, 3).[5] Already at the end of the 1990s, Putnam went even further to assert that the only remaining disparity was that Continental philosophers often lacked clarity or refrained from making explicit arguments (see Putnam 1998, 219).[6]

In recent years, Brandom's contributions have played a significant role in what is perceived as bridging the gap between analytic and Continental philosophy. Notably, Rorty acknowledged Brandom's efforts in overcoming barriers and transitioning analytic philosophy into a Hegelian stage (see Rorty 1997a, 4, or Rockmore 2005, 139).[7] Despite Brandom explicitly stating that he sees himself as an "analytic philosopher" (Williams 2013, 370),[8] he has taken on the ambitious task of integrating Hegel, who was traditionally perceived as an adversary of analytic philosophy, into the analytic debate. His principal argument posits a close alignment between Hegel's objective idealism, specifically his assertion that the world is rational, and his own philosophy of 'conceptual realism'. Brandom contends that understanding mind-independent reality through semantic relations of compatibility and incompatibility—an interpretation of Hegel's notion of determinate negation—can bridge the apparent gap between subjective concepts and the mind-external world. This semantic bridge would then set the stage for an unwavering trajectory of humanity, marked by a dedicated pursuit of improving one's understanding of the mind-external world. Grounded in this newfound understanding, humanity weaves

5 Marshall, David, L. "The Implications of Robert Brandom's Inferentialism for Intellectual History." *History and Theory* 52, 2013: 1–31.

6 Strassfeld, moreover, describes the division as a "gulf of silence" since in the later half of the 20th century both sides have conveniently ignored each other (Strassfeld 2020, 833). Is it possible that an atmosphere of silence has led to Putnam's evaluation that the divide has disappeared?

7 Rorty, Richard. "'Introduction' to Wilfrid Sellars." *Empiricism and the Philosophy of Mind.* Edited by Richard Rorty. Cambridge: Harvard University Press, 1997. Rockmore, Tom. *Hegel, Idealism, and Analytic Philosophy.* New Haven: Yale University Press, 2005.

8 Williams, Jeffrey J., and Robert Brandom. "Inferential Man: An Interview with Robert Brandom." *Symploke* 21 (1), 2013: 367–391. doi:10.5250/symploke.21.1-2.0367.

a harmonious society that cultivates and uplifts individuals, guiding them to evolve into truly "magnanimous" beings (Brandom 2019, 32).[9]

However, Brandom's reading of Hegel's *Phenomenology*, characterized by his unique appropriation of Hegel's terminology, has been a subject of controversy. On the one hand, analytic philosophers still struggle with accepting Hegel's technical and complex prose, which, indeed, often lacks clarity. On the other hand, Continental philosophers express significant concerns about Brandom's rather broad and imprecise interpretation of Hegelian concepts. Pinkard remarks that "the model to which Brandom is actually committed may [...] not be Hegel's at all" (2019, 9),[10] while Houlgate contends that "for the most part, one cannot place one's trust" in Brandom's reading of Hegel's *Phenomenology* (2020).[11] Against these objections, which are more in line with the spirit of Continental philosophy, one could argue that an exhaustive grasp of Hegel's philosophy may not be necessary, provided that Brandom's approach proves effective. As an analytic thinker driven by 'scientific truth', Brandom may therefore not feel compelled to address or seriously consider objections from the Continental tradition regarding potential flaws in his interpretation. Sachs, for example, defends Brandom's appropriation of Hegel, stating that "this is what must happen to make texts come alive in response to a historical situation alien to that in which the text was originally written" (2020).[12] Hence, the subtitle of Brandom's book *A Spirit of Trust—A Reading of Hegel's Phenomenology* suggests that it is one interpretation among many. This makes potential conflicts with other interpretations less relevant.

However, I argue that the issue at hand goes beyond mere interpretation. From my perspective, Hegelian philosophy tends to embrace a historicism that bypasses the challenge of understanding mind-independent reality. This stance sharply contrasts with what I perceive as Brandom's semantic realism, which he calls 'conceptual realism', and which aims to acquire knowledge about reality as such. To put it succinctly, accepting the validity of Brandom's philosophy renders Hegel's historicism redundant. However, considering the challenges

9 Robert Brandom. *A Spirit of Trust: A Reading of Hegel's Phenomenology*. Cambridge: Harvard University Press, 2019.

10 Pinkard, Terry. "A Spirit of Trust: A Reading of Hegel's Phenomenology, by Robert Brandom." *Mind* 129 (515), 2020: 990–999. https://doi.org/10.1093/mind/fzz060.

11 Houlgate, Stephen."Brandom, *A Spirit of Trust: A Reading of Hegel's Phenomenology.*" *Notre Dame Philosophical Reviews,* 2020. https://ndpr.nd.edu/reviews/a-spirit-of-trust-a-reading-of-hegels-phenomenology/.

12 Sachs, Carl B. "Critical notice of Robert Brandom, A Spirit of Trust. A Reading of Hegel's *Phenomenology*." *European Journal of Pragmatism and American Philosophy* XII(2), 2020: 1–7. https://doi.org/10.4000/ejpap.2231.

in verifying Brandom's conceptual realism, Hegel's historicism emerges as the better philosophy. It addresses the complex issue of acquiring knowledge in light of the inherent uncertainty and possible inability of apprehending mind-independent reality. The primary objective of this book is to demonstrate this inherent incompatibility between Brandom's conceptual realism and Hegel's historicist epistemology, ultimately establishing that Hegel's historical account supersedes Brandom's conceptual realism.

A byproduct of this comparison is its capacity to shed some light on the recently obscured divide. By accentuating the incompatible properties of Brandom's conceptual realism and Hegel's historicity, we will unveil a persistent and relevant conflict between Continental and analytic philosophy. While I concede that the continuation of this conflict might not be immediately evident anymore, it would also be premature to assume that established power structures have entirely dissipated. A historical exploration reveals a profound division between two distinctly 'political' camps and I believe that given its former extent it is unlikely to vanish entirely. Strassfeld's research, for instance, provides compelling evidence of the systematic disregard for hiring Continental philosophers within elite departments, resulting in their under-representation within many American philosophy departments (see Strassfeld 2020). As a result, analytic philosophers have coined the views of a whole generation of philosophers and also their views on the Continental tradition. For long, Continental philosophers faced substantial dismissal within analytic circles, resulting in a lack of citations of their work in analytic journals (see Chase and Reynolds 2014, 3–5). The concrete social division has manifested through separate dialogues and institutional barriers, as scholars like Bell have pointed out:

> The division between the traditions has caused dissension, mutual distrust, and institutional barriers to the development of common concerns and problems among working philosophers and so has significantly limited, in many cases, the range and fruitfulness of philosophical discussions and debates.
>
> BELL, CUTROFELLO AND LIVINGSTON 2016, 2

Rockmore arrived at a similar conclusion earlier in 2004:

> Increasing, but selective, attention has been given of late to the relation between analytic and Continental philosophy, which has been marked

> by a century of bad blood. Ignorance, scorn, and lack of attention on both sides have divided them in ways that may or may not be irreparable.
>
> ROCKMORE 2004b, 468[13]

Rockmore, as a scholar of the Continental tradition therefore criticized Putnam's interpretation of a bridged divide as an, in effect, political move suggesting that analytic philosophers could do philosophy better than Continental philosophers (see Rockmore 2004b, 471).[14] Regarding Brandom and the Pittsburgh Hegelians, he objects that a supposed synthesis of two opposed movements simplifies a factual, social conflict that remained for more than fifty years (see Rockmore 2004b, 471).[15]

In addition to discussing the social factuality of the divide, Rockmore also remarks that the divide will persist due to the natural tendency of scientific inquiry to branch into opposing viewpoints: "Philosophy has been divided among proponents of different points of view virtually since the beginning, and that seems unlikely to change whatever happens in analytic philosophy" (2001, 369).[16] Taking this perspective into consideration, the answer to Putnam's question, "why can we not just be 'philosophers' without an adjective?" may even lie in the fact that the conflicting and passionate positions are integral to the emergence of knowledge (see Putnam 1998, 221). Moreover, Rockmore's suggestion that the divide may stem from a deeper philosophical

13 Rockmore, Tom. 2004b. "On the Structure of Twentieth-Century Philosophy." Metaphilosophy 35 (4): 466–478. http://www.jstor.org/stable/24439712.

14 As also Rorty notes: "they identify philosophical ability with argumentative skill and notice that there is nothing they would consider an argument in the bulk of Heidegger or Foucault, suggest that these must be people who tried to be philosophers and failed—incompetent philosopher" (Rorty, Richard. "Philosophy in America Today." *The American Scholar* 51 (2), 1982: 183–220, 185).

15 Against the more 'book-based' approach of analyzing the divide Strassfeld writes: "This change was not, primarily, the result of conscious efforts to identify or create an 'analytic' philosophical movement. Rather, language evolved to describe the material reality of a broad discursive community that emerged from the practical and professional activities of logicians and linguistic philosophers" (2020, 837). Rorty indicates similar problems when it is about academic struggles for positions: "Such disputes over money and power produce a lot of ad hoc rhetoric about the nature of the discipline, rhetoric that an experienced dean will automatically tune out. Such disputes only become dangerous when one side wants to say that the material taught by the other side should not be taught at all [...] I have heard analytic philosophers get furious at comparative literature departments for trespassing on philosophical turf by teaching Nietzsche and Derrida, and doubly furious at the suggestion that they might teach it themselves. Conversely, I have heard fans of Continental philosophy be obnoxious about the 'mere logic-chopping' with which their analytic colleagues waste students' time and dehydrate their minds" (Rorty 1982a, 196).

16 Rockmore, Tom. "Analytic Philosophy and the Hegelian Turn." *The Review of Metaphysics* 55 (2), 2001: 339–370, http://www.jstor.org/stable/20131711.

underpinning is also a notion that Rorty alludes to: "the heavy breathing on both sides about the immorality or stupidity of the opposition signals passions that academic power struggles cannot fully explain" (Rorty 1982a, 199). In other words, the divide may not merely be a matter of stylistic differences or academic power struggles, but also the result of opposing viewpoints forged through intense discourse and shaped by passionate, experiencing, and debating individuals. It may be the outcome of a dialectical process within history, as discussed by Hegel, and a result of our human nature to interact, debate, and thereby produce history. Consequently, it may not simply disappear.

Due to various reasons, the insistence on the concept of a bridged divide may inadvertently obscure social facts and oversimplify the debate. For instance, the analytic tradition tends to favor an ahistorical approach, potentially undermining the careful textual examination practiced in the Continental tradition. This becomes evident, especially when we observe that Brandom's book largely ignores the extensive tradition of secondary commentary on Hegel (see Pinkard 2020, 1). Considering the presence of these two distinct camps of philosophers is, therefore, not only helpful but also essential in navigating the complexities of the discourse. Regarding a possibly underlying passionate and dialectical reason for developing oppositions, I am, however, not going so far as to state that we "cannot understand either analytic or Continental Philosophy without its 'other'" (Chase and Reynolds 2014, 9). I do not claim that Continental and analytic philosophy are both two sides of one dialectical coin. However, I believe that when it comes to Brandom and Hegel it can still be meaningful to examine their historical interaction as adversaries from two different traditions. This entails reconstructing their standards while considering their historical contexts.[17]

Regarding the frame of the divide, the book therefore analyzes two very different, but interlinked philosophies. I reconstruct Brandom's philosophy as the continuation of Sellars' anti-atomist approach, which opposes the anti-Hegelian early analytic philosophy. On the other hand, I reconstruct Hegelian philosophy as defending an opposing perspective centered on the mind-internal, historical relativism emerging from the Idealist tradition. If my reconstruction is feasible, it will become evident that Hegel's historicist relativism is incompatible with Brandom's conceptual realism. Moreover, this incompatibility between Brandom and Hegel does not stem from Hegel being

17 Of course, such systematic relations are multi-dimensional, and oppositions are also developed inside of these disciplines. For this reason, I am not reducing the essence of Continental and analytic philosophy to two systematic parts of a whole, but pointing out a weaker systematic relation resulting from their confrontation.

a metaphysician with an implicit "onto-theology," as many critics have argued (see, for example, DeVries 2021a).[18] Instead, it arises from Hegel's philosophy advocating for a historical, relativist empiricism that permits only dialectical speculation on the concept of a mind-independent reality. In essence, Hegel's Continental approach emerges as a direct competitor to Brandom's predominantly analytic conceptual realism. In the following paragraphs, I seek to clarify the context of this abstracted opposition between Continental historicity and analytic conceptual realism.

Philosophical perspectives are influenced and transformed throughout history as they come into contact with opposing positions. Proponents of each camp often seek to distance themselves from the other, resulting in alterations to their philosophical frameworks. Regarding the conflict between Brandom and Hegel, we can observe this phenomenon in the early days of the analytic tradition, exemplified by Russell's anti-Hegelian movement. Russell famously declared, "I began to believe everything the Hegelians disbelieved" (1995, 48).[19] In the ensuing debate, we witness the "reassertion of atomism" and an explicit commitment to a materialist realism and scientism (DeVries 2017, 2).[20] At the core of the new philosophy lies the rejection of what is perceived as Hegelian, unscientific, "mentalistic vocabulary" (DeVries 2017, 2). Rockmore's analysis suggests that as a result analytic philosophy is fundamentally shaped by its opposition to idealism:

> analytic philosophy tends to define itself through its opposition to idealism, where the latter is understood as whatever analytic thinkers reject. This longstanding negative attitude toward idealism still exists for many analytic philosophers. But it has been partly defused for others through the recent selective nascent analytic turn (or return) to Hegel.
>
> ROCKMORE 2004b, 472

The beginning of analytic philosophy is notably marked by the strong repudiation of Hegel's idealism.[21] This initial rejection laid the foundation for a

18 DeVries, Willem A. 2021a. "Hegel Today." *Aeon*. Edited by Sam Dresser. https://aeon.co/essays/how-the-anglophone-world-is-rediscovering-hegels-philosophy.

19 Russell, Bertrand. *My Philosophical Development*. London: Allen and Unwin, 1995.

20 DeVries, Willem A. 2017. "Hegel's Revival in Analytic Philosophy." The Oxford Handbook of Hegel. Edited by Dean Moyar. DOI: 10.1093/oxfordhb/9780199355228.013.35.

21 Much attention has been given to explain the divide in terms of the "encounter between Carnap, Cassirer and Heidegger" (Strassfeld 2020, 835). Strassfeld criticizes that this encounter has placed too much emphasis on the Continental side and has not yet investigated the role of the analytic side (see Strassfeld 2020, 835). He looks at the "vectors of

divide that led to a problematic understanding of what idealism truly entails, shaping both our perception of Hegelian philosophy and the trajectory of analytic thought. During its formative stages, analytic philosophy championed an anti-holistic approach, striving to ascertain immediate truths as reflections of a mind-external reality. While subsequent discussions quickly revealed the fallacy of this perspective, the problematic rejection of Hegel's idealism persisted. According to Russell, idealism implies that only the mind exists, a definition that has become the standard interpretation found in dictionaries and scholarly debates (see, for example, Irvine 2015).

Of course, Hegel's idealism does not assert that only the mind exists. Instead, Hegel's *Phenomenology of Spirit*, for example, challenges our assumption of an immediate connection between the mind and a mind-external reality. In the analytic tradition, Sellars was arguably the first to rediscover this Hegelian perspective and introduced it into the philosophical discourse. In opposition to the so-called myth of the given—which represents analytic philosophy's "regulative idea"[22] of an atomistic realism—Sellars initiated the reevaluation of Hegel within neo-pragmatist circles (DeVries 2021a). Now, Hegel emerged as a pragmatic, holistic, and social thinker who espoused a collectively shared non-metaphysical rationality.

However, in Sellars' quest to overcome idealism (see Sellars 1968, 143), the significant relationship between Hegel's philosophy and idealism, as well as historicity, was neglected.[23] Rockmore criticizes this partial appropriation of Hegel:

> Though Hegel is a leading German idealist, analytic philosophers now turning to Hegel routinely draw a tacit distinction between Hegel and idealism. Everything happens as if it were possible to appropriate Hegel for analytic concerns while simply bracketing his idealism. Though they

influence in the innumerable points of contact between philosophers and the institutions in which their labor is situated" (Strassfeld 2020, 835).

22 According to DeVries, analytic philosophy can be more or less associated with "foundationalism, individualism, reductionism, and materialist realism" (see DeVries 2017, 3). He further thinks that the "revival of Hegelian thought" can only be achieved if we discard the "atomistic foundationalism" which, according to him, is the goal of Sellars (see DeVries 2021a).

23 "But *no* picture of the world contains as *such* mentalistic expressions functioning *as such*. The indispensibility and logical irreducibility of mentalistic discourse is compatible with the idea that in *this* sense *there are no mental acts*. Though full of important insights, Idealism is, therefore, radically false" (Sellars, Wilfrid. *Science and Metaphysics—Variations on Kantian Themes*. New York: Humanities Press, 1968, 143).

> are starting to come to grips with Hegel, to the best of my knowledge none of them has yet found a way to come to grips with his idealist commitment or with idealism in general [...] The nascent Hegelian turn in analytic philosophy is distinguished by its utter disregard for the deeply idealist dimension of Hegel's theory.
>
> ROCKMORE 2004b, 472

Rockmore argues that the reintroduction of Hegel into philosophical discourse can only be truly beneficial if we thoroughly comprehend his idealism, as it elucidates crucial aspects of our relationship to the concept of a mind-independent reality.

At this juncture, one could argue that Brandom has made an attempt to incorporate Hegelian idealism from *The Phenomenology of Spirit* into his conceptual realism. In his work *A Spirit of Trust*, Brandom distinguishes Hegel's idealism from the 'bad' idealists (see Brandom 2019, 52). There, as already mentioned, he portrays Hegel's objective idealism as a form of his conceptual realism. Nevertheless, Brandom's rediscovery of Hegel is still driven by the initial conflict within analytic philosophy that resulted in the rejection of Hegel. In alignment with Sellars' anti-idealism, Brandom is concerned with the question of how to access a mind-independent reality as a standard for discourse. On the contrary, I interpret Hegel's idealism as a relativist historicism that is fundamentally incompatible with this standard. In other words, while Brandom follows the early analytic and later Sellarsian approach of seeking access to mind-external reality, Hegel is primarily interested in the idealist inquiry of how to structure human historical experiences without assuming such access.

Early in the debate, Habermas brings Brandom's reliance on this mind-external, ahistorical reality to the fore. In *Truth and Justification* [*Wahrheit und Rechtfertigung* (1999)], he explicitly rejects Brandom's conceptual realism of *Making It Explicit* (1994). Even though Brandom's early work does not mention Hegel by name, it represents, according to Habermas, a false return to a 'metaphysical' Hegel. His major objection against Brandom's Hegelianism consists in the fact that the concept of the mind-independent world remains a self-produced, human concept. It therefore cannot be used as a formal mind-external standard that guides our discourses.

While I agree with Habermas' critique of Brandom's position and his, in my view, return to relativism, there is a broader theme that drives my discussion of Habermas' stance. In his criticism of Brandom, Habermas relies on a common interpretation of Hegel, one that I find concerning. Similar to Brandom, he interprets Hegel as advocating for an ahistorical connection between our knowledge claims and mind-external reality. I, therefore, argue that both

Habermas and Brandom overlook or distort Hegel's concept of what I refer to as 'historicist systematicity'. By adopting an approach rooted in historicist systematicity, which aligns closely with Hegel's idealism, the concept of access to a mind-external reality becomes unnecessary.

What do I mean by 'historicist systematicity'? The concept of historicist systematicity can be seen as a key differentiating factor between Continental and analytic philosophy. Franks, for example, comes to the following conclusion: "the German idealist conception of systematicity and its value stands between the Anglo-American and Continental traditions, which still, for the most part, pass each other by in mutual incomprehension" (Franks 2005, 3).[24] Franks further posits that the analytic position retains its foundational stance of rejecting idealism as a closed, metaphysical system (Franks 2005, 3). However, the emphasis within Continental philosophy is not on the existence of a single system. Instead, it concerns itself with the historical failures associated with the traditional conception of *the* system (see Franks 2005, 4). Consequently, Continental philosophy has progressed towards developing an understanding of systematicity that incorporates an openness to history. Hegel, often criticized for endorsing an unwarranted and "grandiose" system (DeVries 2021a), actually formulated and employed an open historicist position already within the *Phenomenology of Spirit*. As I will show, he introduces the notion of a system open toward historical change, which, however, gives rise to a contentious form of global relativism. This historical systematicity is the major reason for why I believe his philosophy is incompatible with the ahistorical, conceptual realism advocated by Brandom, which asserts that the structure of our discourses is guided by their relation to a mind-independent reality.

Hegel's account is often, in my view, falsely portrayed as a teleological system that presents an ahistorical categorical structure of rationality, leading to an implausible superstructure. Habermas, for instance, follows this interpretation. However, I will present a reconstruction of Hegel as a historical thinker who believes that all our cognitive claims are limited to their historical context. For instance, in Hegel's view, systematicity means to reconcile conflicting positions through a higher unification. It involves the development of historically emerging positions towards a system based on an anthropological understanding of subjects immersed in their history. This process of systematization initiates at the level of natural consciousness and extends across all

24 See Franks, Paul W. *All or Nothing: Systematicity, Transcendental Arguments, and Skepticism in German Idealism*. Cambridge: Harvard University Press, 2005, 3.

levels of cognition. In essence, systematization is steered by the experiences and reflections of historically situated individuals who assert knowledge claims and thereby posit concepts. According to Hegel's model, claims evolve progressively into extreme theoretical positions and are consistently reconciled into new unities. The overarching objective is to achieve a higher degree of systematization of these positions. Ultimately, the series of historically limited, but interconnected systems resembles a movement towards truth. In my interpretation of Hegel, the essence of this systematization is internally reconstructed through reflecting on human experiences within historically-limited, yet evolving societies and cultures. I will analyze Hegel's *Phenomenology* from this perspective of an immersed, historically limited subject striving for systematization towards an ideal unity of truth.

When considering the process toward higher systematization, two interconnected questions arise: firstly, can we conceive of this systematic process solely in formal terms? And secondly, is it possible to achieve a complete and closed unification within a system? Regarding the first question, it is challenging to establish systematicity as a definitive intra-theoretical criterion within our epistemological discourses in the present historical context. Any attempt to provide a formal account of systematicity has become so far entangled with external factors and complexities. Hegel, instead, regards systematicity as a primarily historical description of how consciousness operates. Therefore, systematicity serves as a criterion that manifests itself in relation to various historical contents and within the context of its current historical moment. Systematicity is comprehended as a concept rooted in history rather than being approached solely from a formal perspective. Regarding the second question, if Hegel's account of systematization is intimately connected to history, it becomes challenging for Hegel to uphold the notion of a final form of the system unless we have arrived at the end of history. This viewpoint aligns with Rockmore's perspective on Hegel's historicism: "If Hegel is correct, the commitment to contextualism already entails a commitment to a historical conception of knowledge, hence to historicism, that is to a view about the intrinsically historical character of all of all first-order cognitive claims" (Rockmore 2005, 193). In other words, the contentious proposition is that Hegel suggests that all our knowledge claims are shaped by historical processes and cannot be comprehended in isolation from history. There is no definitive or conclusive statement that marks the end of debate in the current historical moment. As a result, Hegel presents a concept of systematic, historical relativism wherein there is no ultimate or final system.

I acknowledge that this interpretation is subject to controversy and cannot be advocated as the sole interpretation of Hegel. In his book *Hegel's Idealism*

(1989), Pippin highlights the commonly held belief that "no one really knows what [... Hegel] was talking about" (Pippin 1989, 3).[25] Given that this statement has not lost its historical truth, the goal is, therefore, to present my interpretation as a viable option for reading Hegel, especially within the spectrum of the recent non-metaphysical readings of Hegel. Nonetheless, this interpretation sharply contrasts with Brandom's perspective, which overlooks the dialectical, historical, and relativistic dimensions inherent in Hegel's work. Beyond mere differences in interpretation, I contend that Hegel, when approached through a relativistic lens, provides a more encompassing comprehension of the epistemological problem. His perspective offers a more viable relativistic solution that does not hinge on an explicit connection between mind and reality.

However, my perspective on Hegel is not only controversial because I argue that Hegel espouses a form of relativism, but also because relativism is commonly believed to render knowledge impossible and fails to justify its own position, as it denies knowledge claims altogether.[26] Contrary to this, I propose that Hegel's historical relativism is compatible with a particular theory of truth, offering a modified version that addresses the standard objections against relativism. The typical argument against relativism goes as follows: if we consider the claim 'everything is relative', there are two common strategies employed to undermine the relativist position. The first strategy involves a strong relativism that asserts the truth of the claim itself, namely, that everything is relative. However, this approach leads to a performative self-contradiction, as the content of the claim contradicts its own performance. Consequently, a second, weaker version of relativism emerges, which seeks to avoid this performative contradiction. In this case, the relativist does not assert the absolute truth that there is no truth, but rather claims, 'according to my perspective, there is no truth in an absolute sense'. However, this position leaves open a second strategy for an interlocutor to claim that his perspective is different in that there is such thing as truth. In both these cases, relativism seems difficult to defend.

The claims of the historical relativism that I am defending are different. For, if I claim that 'everything is historically relative', then somebody could demonstrate that my claim is itself historically relative. In this way, my claim, however, does not stand in contradiction to my performance. Thus, avoiding the first objection to relativism. Now, while I claim that everything is historically

25 Pippin, Robert. *Hegel's Idealism. The Satisfactions of Self-Consciousness*. Cambridge: Cambridge University Press, 1989.

26 As Field observes: "But the term 'relativism' has had the misfortune of being defined by its opponents" (Field, Hartry. "Epistemology without metaphysics." *Philosophical Studies* 143 (2), 2009: 249–290, 255. doi:10.1007/s11098-009-9338-1).

relative, although my position depends on my history, it still, nevertheless, could be true. One could object that my claim excludes meaningful claims, but, in fact, it is not excluded that my historicist claim could be true. I only deny that so far we can identify it as independent of history. Independent of this history, however, it could be true. So, in this case I am open to a discussion on rational standards and include the possibility of falsification. This is the decisive difference between mere relativism, which excludes any possible opening toward truth, and historicist relativism. In other words, historicism is in theory compatible with 'the aim of truth' (as we will see, this is a major formulation of Hegel).

Margolis points to the openness of historicism: "Historicism, in short, is itself a universalized doctrine found under quite contingent, particular circumstances to be distinctly compelling but not conceptually inescapable at every possible price" (Margolis 2007, 310).[27] I interpret this to mean that historicism must acknowledge the potential for an openness to truth. Consequently, historically limited systematization can assert its validity as potentially true. It is not solely concerned with the local arrangement of certain knowledge fragments, which we sometimes deem valid in certain fields, but rather, it can establish a connection to a broader truth. A historicist relativism can thus justify itself without encountering performative contradiction or losing relevance. 'Historical systematization' must then be understood as the extended pursuit of a unified theory encompassing knowledge, ethics, art, religion, and philosophy. Yet, although historicist relativism endeavors to grasp the entirety and therefore exhibits a holistic approach, it remains confined within the boundaries of its historical contexts so far and equally acknowledges them.

Margolis also hints at various historicisms attempting to avoid relativism: "In fact, it is a characteristic of numerous historicist theories that they end by embracing some form of universalism. The reason is simply that historicists are often motivated by the wish to avoid relativism or the appearance of relativism" (Margolis 2007, 40). Here, I want to emphasize again that I am not denying the relationship between historicism and relativism. However, I also do not exclude the compatibility of historicism and truth. I cannot say 'everything is historical and I know this as a truth', but I agree that a branch of historicism might end at some node because we have found a truth. Or, it could also be that we find an ultimate truth through investigating history and following its historicist thread. The task of historicism is therefore not to deny

27 Margolis, Joseph. *Pragmatism without Foundations: Reconciling Realism and Relativism.* London: Continuum, 2007.

the possibility of truth, but to test historical positions with regard to their standards. My controversial claim is that the relativist historicism I argue for aligns with Hegel's view, and I intend to demonstrate this.

Given the significant differences between Hegel's historicism and Brandom's ahistoric, conceptual realism, I agree with Rockmore that "the distinctions between analytic and Continental philosophy are real and persistent and cannot obviously be bridged merely through continuing the debate" (Rockmore 2004b, 474). Rockmore, moreover, excludes that the link between analytic and Continental Philosophy can be established by the fundamental rule of logic that is often central to analytic approaches. This is also because the Hegelian logic is distinct from the ordinary, logical programs (see Rockmore 2004b, 474). Rockmore therefore also rejects a Brandomian notion of the mind-external whole understood as a semantic, conceptual net of implicit logical relations. Instead of grasping reality as it is, Rockmore puts emphasis on constructivism:

> The promise of the new century does not lie in continuing the fruitless effort, which stretches back in the Western discussion at least until Plato, to know the way the world is. As Kant suggests, it lies rather in further working out a constructivist approach to knowledge.
>
> ROCKMORE 2004b, 477

It was often assumed that this constructivism must be the attempt to find absolute, objective criteria for how we construct subjective reality. In fact, this was also the early discourse-theoretical program of Habermas and Apel.[28] However, we can understand and argue for a constructivism that is grounded on the idea of reconstructing our own history, which is the relative result of our societal collaboration and our own actions. It therefore still defends criteria that serve as objective means in our historical moment, but that are not

28 Habermas also criticizes that the procedural accounts of truth, in particular, to deduce the ultimate conditions for discourses in order to get closer to reality, have been unsuccessful. Apel, for example, attempted to point out the undeniable, preceding conditions for participating in any discourse. Such transcendental conditions, however, are only recognized for specific situations according to Habermas. Habermas ultimately rejects this view in *Erläuterungen zur Diskursethik.* For Apel, this means the dissolution of discourse ethic [Auflösung der Diskursethik] (Apel 1988, 727). With regard to an ultimate foundation [Letztbegründung], Habermas comes to the conclusion that the foundation, in the form of a substantial derivation from paradigmatic norms of actions, is not only impossible, but also unnecessary (see Habermas 1991, 195). Habermas, Jürgen. 1991. *Erläuterungen zur Diskursethik.* Frankfurt: Suhrkamp. Apel, Karl-Otto. *Diskurs und Verantwortung. Das Problem des Übergangs zur postkonventionellen Moral.* Frankfurt: Suhrkamp, 1988.

absolute in the strict sense. They are absolute with respect to what we have historically experienced as finite human beings. This historicist constructivism is relativist but not mere relativism. The advantage of a historicist over a mere relativist approach is that it allows us to develop and justify knowledge claims that are better than other knowledge claims. These knowledge claims are based on how we have historically come to understand and make objects. It can test absolute truth claims and allows us to falsify them with regard to our further historical development. Most importantly, however, it does not exclude a possible opening towards an everlasting truth. The major objective of this book is therefore to develop this relativist historicism in its roots for the Hegelian philosophy, particularly with regard to the *Phenomenology of Spirit*.

The relativist-historicist interpretation of Hegel's *Phenomenology* that I propose will occur to many as controversial since the *Phenomenology* should, as Hegel even states himself, be regarded as the ladder to the *Logic*. In the *Logic*, Hegel seemingly articulates the ahistorical core of his philosophy. Given this ahistoric core, there will be strong objections against my reading of Hegel as a global relativist. Nevertheless, some will also say that my relativist-historicist interpretation of Hegel knocks on open doors. There are, indeed, many philosophers who suggest a similar interpretation of Hegel. Jaeschke, Rockmore, Sedgwick, Žižek, and, as I could read in his critiques of Brandom, Pinkard make strong arguments that support a relativist-historicist reading of Hegel, some of them even with regard to Hegel's *Logic*.[29] In addition to discussing these accounts, I will explain my proposed interpretation in the light of Hegel's later *Lectures on the History of Philosophy*. My goal, however, is not to weigh and balance all the counter-arguments against and for my reading of Hegel but first of all to ground the relativist-historicist reading with regard to Hegel's *Phenomenology*. It will unveil a profound contrast between Brandom's ahistoric conceptual realism and Hegel's emphasis on historicity, with the expectation of pointing towards a more tenable relativist direction.

The book is structured in three chapters. First, it consists of a reconstruction of Brandom's basic account as a result of analytic interpretations of Hegel's idealism. It will focus on Brandom's strong reliance on mind-external reality as an

29 I do not discuss Pippin here since I do not understand his recent book *Hegel's Realm of Shadows*. It seems to have a stronger stance on anti-relativist elements in Hegel's *Logic*. As DeVries points out, however, it has been Pippin and Pinkard who strengthened the "non-metaphysical" interpretation of Hegel. They argue that Hegel criticizes "rationalistic metaphysics" which employ "fundamental truths" (See DeVries 2021a). Rather, it is about a mind-internal investigation of "human rationality" that develops a historical and "categorial scheme" in order to interpret phenomena (see DeVries 2021a).

ahistorical standard and conclude with a discussion of Brandom's recent book on Hegel: *A Spirit of Trust* (2019). Second, it will follow Habermas' criticism of Brandom's conception on the one hand. But also discuss Habermas' interpretation of Hegel's concept of the absolute as ahistoric, on the other hand. Third, the book will deliver what I perceive as a corrected view on Hegel's relativist historicism. Overall, this book is of course not an attempt to bridge the divide, but to present a different direction for philosophy that Hegel's relativist historicism, as I understand it, can still offer to us.[30]

30 A note has to be made on the extensive references to Rockmore's philosophical assessment of the relevant philosophical figures. Rockmore does not develop his own philosophy, but is for the most part a historian of philosophy. His own view, though never explicitly expressed as his own position, consists in the idea that philosophy is a result of traditions. Besides Hegel, his work has the largest impact on this book. I follow his thesis that Hegel develops a phenomenology as well as I share his views on the analytic tradition accessing Hegel. Since he, however, does not explicitly develop his own philosophy, I will not refer to his philosophy directly. Rather, he will occur frequently in the role of a commentator.

CHAPTER 2

Brandom's Conceptual Realism and Two Interpretations of Hegel

In this chapter, my goal is to discuss Brandom's conceptual realism as an attempt to pragmatically integrate semantics,[1] while showcasing that this attempt itself stems from two problematic interpretations of Hegel's idealism in analytic philosophy. In the first part, I will trace the roots of Brandom's conceptual realism back to Russell's and Moore's questionable rejection of Hegel as an anti-realist. I will argue that this rejection, which I find flawed, leads to Sellars' interpretation of Hegel as a social thinker. After providing this historical introduction, I will address Brandom's conceptual realism as a response to traditional problems in analytic philosophy and present his specific approach to integrating semantics pragmatically. Overall, I argue that Brandom's project is based on what I perceive as a problematic relation to mind-external realism.

1 From Hegel to Analytic Philosophy to Pragmatic Contextualism

The analytic view on Hegel and its later implementation in the discourse of Brandom's philosophy is based on a major misinterpretation of Hegel's philosophy in the late 19th century. Due to Hegel's late rise in British thought, the rejection of Hegelianism inaugurated by Russell and Moore, occurs about fifty years later than in Germany and thus gives British philosophy (and more widely non-Continental philosophy) a different relationship to Hegel. While Hegelian philosophy had already gone through a cycle of rejection on the continent and its influence had already declined, Hegelian idealism emerged in England and Scotland, and was also popular in the United States in the latter half of the 19th century. The late popularity of Hegel in the English tradition was linked to the late translations of Hegel's works. So, for example, William Wallace's translation of Hegel's *Encyclopedia* was published only in 1874, while the first translational work began in the 1860s (see Mander 2011, 40).[2] William

1 The term 'pragmatic integration of a reference-semantic' was originally coined by Edmund Braun (see Braun, Edmund. *Der Paradigmenwechsel in der Sprachphilosophie.* Wissenschaftliche Buchgesellschaft Darmstadt, 1996).

2 Mander, W.J. *British Idealism: A History*. Oxford: Oxford University Press, 2011.

 DOI:10.1163/9789004714809_003

Wallace was fascinated that Hegel could replicate the concept of the evolution of species as the evolution of thoughts. However, Wallace's Darwinian interest in the "actual historical development of things" (Mander 2011, 43) already foreshadows the main competitor of Hegel's project, the natural sciences. Wallace explicitly rejects Hegel in favor of the natural sciences. While on the continent the influence of Hegel's philosophy had already disappeared as a result of the struggle of materialism [*Materialismusstreit*], in British philosophy, the rejection of Hegelian philosophy as 'unscientific' had just begun.

Rockmore delineates three distinct philosophical perspectives that arose in response to Hegel during the latter part of the 19th century, each offering its unique interpretation of Hegelian ideas: "American pragmatism,[3] analytic philosophy, and [...] the phenomenological movement" (Rockmore 2001, 340).

The phenomenological movement, to take the last first, stands for the new Continental philosophies but does not necessarily relate to Husserl's foundationalist phenomenology that seeks to identify evident grounds after a phenomenological reduction. If it is read independently of Husserl's interpretation, the phenomenological movement rather reinterprets what empiricism means by reanalyzing our experiences without the assumption of a noumenon. As Rockmore often argues, the phenomenological movement can be interpreted as further developing Hegelian ideas. The phenomenological movement occurred after Hegelian philosophy had declined and during the rise of empirical sciences. The later Continental reinvigoration of metaphysical topics constitutes a major element of the divide since the early analytic philosophy claims to reject Hegel on the grounds of a scientific realism that discards metaphysics.

American pragmatists, such as Peirce, claim that their philosophy is intrinsically related to Hegel. Although, in my view, their philosophy with its focus on action might be interpreted as genuinely Hegelian, Brandom's normative pragmatics is more interested in the semantic question of how context influences claims. For this reason, my investigation does not need to explore the American pragmatists to understand Brandom's particular appropriation of Hegel. It is the analytic tradition, from which Brandom's philosophy emerges, which constitutes a strong contrast to the phenomenological movement. With regard to analytic philosophy and its relation to Hegel, Rockmore comes up with the important thesis that there are two historical misunderstandings of Hegel's philosophy:

3 In the United States Josiah Royce, Gustavus Watts Cunningham, Charles Sanders Peirce and John Dewey were influenced by Hegelian ideas (see DeVries 2021a).

> If I am right, Hegel turns out to be a key but mainly unacknowledged figure in analytic philosophy, from whom it departs and to which it is now returning, but which it has twice misunderstood. The first, highly productive misunderstanding was one ingredient in the rise of analytic philosophy. It remains to be seen whether the second misunderstanding, which is only now taking shape, will prove as productive in the further evolution of analytic philosophy.
>
> ROCKMORE 2001, 340

The central focus that needs to be analyzed in order to grasp Brandom's analytic interpretation of Hegel lies in this double misunderstanding of Hegel and the pivotal moments of the reaction to Hegel within analytic philosophy. Firstly, analytic philosophy misinterpreted Hegel as denying the existence of the external world. However, I argue that Hegel does not deny the existence of mind-external reality. Secondly, when analytic philosophy later revisited Hegel and embraced more anti-realist arguments, it fostered a perception of Hegel solely as a social thinker, disregarding his fundamental concept of historicity. Brandom's conceptual realism, in a broad sense, can be traced back to Russell's and Moore's problematic rejection of Hegel as an anti-realist, as well as analytic philosophy's subsequent return to Hegel as a social contextualist. To truly understand Brandom's analytic Hegelianism, it is essential to delve deeper into the original analytic interpretation of Hegel and its evolution towards pragmatic contextualism.

1.1 *The Creation Myth of Analytic Philosophy—The Rejection of Hegel*

Russell, as a founder of the analytic tradition, presented the 'birth document' that explicitly denounces Hegel's ideas. When Russell 'escaped' from the, in his words, "hot-house on to a wind-swept headland"[4] he meant to reject the core-conceptions of the idealist framework (see Russell 1995, 48).[5] As a part of his creation myth of analytic philosophy, Russell focused on the complete opposite of everything that could be Hegelian: "I began to believe everything the Hegelians disbelieved" (Russell 1995, 48).[6] In particular, Russell assumed

4 Russell, Bertrand. *My Philosophical Development*. London: Routledge, 1995.

5 Rockmore makes various remarks on the definition of *Idealism*. While the concept remains elusive within the German Idealist tradition, it is typically understood as a "binary opposition" by British thinkers and Marxists, "which opposes idealism to something else, such as materialism, realism, naturalism, or even empiricism" (Rockmore 2001, 342).

6 Redding writes on this aspect: "As he [Russell] tells it, it was his work on Leibniz that had led him to the topic of relations and there he discovered a thesis at the heart not only of Leibniz's metaphysics but also of the 'systems of Spinoza, Hegel and Bradley'" (Redding, Paul.

that the major mistake of Idealism was hidden in the Leibnizian predicate logic: "every proposition attributes a predicate to a subject and (what seemed to him almost the same thing) that every fact consists of a substance having a property" (Russell 1995, 48). Predicate logic originated from the observation that every substance is related to a predicate. Yet, Leibniz interpreted this relationship as indicating a proximity to Aristotelian syllogistic reasoning, thereby implying a fundamental reliance on substances. For Russell, this was the main mistake of Idealism (see Redding 2007, 3). Russell further maintained the assumption that Hegel still depended on these antiquated forms of logic, even though Hegel himself aimed to rejuvenate logic, albeit not in a manner aligned with Russell's understanding.

Moreover, in his examination of Hegel, Russell does not make a clear distinction between Bradley's perspective and Hegel's. Russell states: "Mr Bradley has worked out a theory according to which, in all judgment, we are ascribing a predicate to Reality as a whole; and this theory is derived from Hegel" (Russell 1914, 38).[7] Hegel, as I think should be understood, does not imply such an ontological monism, according to which there is only one ultimate substance. Instead, Hegel discusses the question of unity from an epistemological standpoint. Russell includes a short discussion on this aspect (see Russell 1914, 37–38), yet he assumes that Hegel's logic is identical to metaphysics, while metaphysics for Russell represents the universe as it is. Beyond this assumption that Hegel is a metaphysical realist of the mind, Russell proceeds with a discussion of the traditional status of logic in Hegel:

> Now the traditional logic holds that every proposition ascribes a predicate to a subject, and from this, it easily follows that there can be only one subject, the Absolute, for if there were two, the proposition that there were two would not ascribe a predicate to either. Thus Hegel's doctrine, that philosophical propositions must be of the form, "the Absolute is such-and-such," depends upon the traditional belief in the universality of the subject- predicate form.
>
> RUSSELL 1914, 38

In this citation, Russell aims to critique Hegel's all-encompassing notion of the Absolute, which is based on the traditional predicate-subject structure of logic that Russell rejects. Instead, Russell puts forth a view centered around

Analytic Philosophy and the Return of Hegelian Thought. Cambridge: Cambridge University Press, 2007, 8).

7 Russell, Bertrand. *Our Knowledge of the External World.* Chicago: Open Court, 1914.

logical atoms as the fundamental components that establish the meaning of a sentence. Unlike Hegel's presumed conception of the world as a unified whole, where all things are ascribed as predicates to a single absolute subject (the "bowl of jelly"), Russell proposes an alternative perspective known as the "bucket of shot" view (Redding, 2007, 4). According to this view, subjects possess predicates as we describe them. Russell advocates for an anti-monist and anti-holistic approach, replacing Hegel's internal relations between subjects and predicates with external relations that are devoid of an all-encompassing subject (see Redding, 2007, 4).

Russell's critique of ontological monism does not inherently justify his leap to ontological pluralism, as this move may entail an unjustified dichotomy. It is important to note that Russell acknowledged the lack of demonstration or evidence for ontological pluralism. Nevertheless, likely influenced by Ockham's razor, Russell leaned towards ontological pluralism as a preferable stance over what he perceived as Hegelian ontological monism— the notion that all of reality is the super-subject of an absolute mind. However, in Russell's interpretation of Hegel as an ontological monist, he overlooks Hegel's stance on epistemology, which casts doubt on our ability to fully grasp a reality independent of the mind. Furthermore, Russell later developed the concept of neutral monism, which presents an intriguing alternative to the challenges posed by distinctions between mind and matter. However, this viewpoint still neglects Hegel's dialectical development of these concepts and fails to engage with the question of history.

Russell's interpretation of the Hegelian tradition had a significant influence on analytic philosophy. This influence becomes apparent when we observe that Russell's definition of Idealism resurfaces in the Stanford Dictionary, where it is articulated as follows: "Like idealism (the view that nothing exists but the mental) and physicalism (the view that nothing exists but the physical), neutral monism rejects dualism (the view that there exist distinct mental and physical substances)" (see Irvine 2015).[8] However, it is important to note that German Idealism does not propose that everything is solely mental. Instead, it presents an epistemological standpoint in which philosophers question our knowledge of the world through representations. The general belief is that this inquiry begins subjectively and aims to establish objectivity. This rough characterization, influenced by Kant's Copernican turn, does not imply that the world is exclusively mental.

8 Irvine, Andrew David. "Bertrand Russell." *The Stanford Encyclopedia of Philosophy*. Edited by Edward N. Zalta. 2015, accessed on July, 5 2017, https://plato.stanford.edu/archives/win2015/entries/russell.

It is crucial to highlight that Russell's perspective, considering idealism as a theory that posits reality as purely mental, greatly influenced analytic philosophy. However, this viewpoint represents a problematic interpretation of Hegel. If we follow non-metaphysical interpretations, Hegel's primary concern lies in the realm of epistemology—how to justify knowledge—rather than ontology—the nature of the world itself (see Rockmore 2005, 52). Questioning how we come to know the world is a less contentious matter than questioning the existence of an external world. However, due to Russell's interpretations of idealism, the question of the external world gained prominence in early analytic philosophy.

1.2 *The Problem of the External World in Analytic Philosophy*

In his analysis of the early analytic tradition, Rockmore specifically examines Moore's perception of the British idealists. Moore viewed the British idealists primarily as opponents of the British empiricists, who asserted that we discover reality rather than construct it (see Rockmore 2001, 344). However, Rockmore criticizes Moore for making several conflations, similar to Russell, when assuming that "modern idealists" assert that the "universe is spiritual" (Rockmore 2001, 345): [...] it simply ignores Hegel and other German idealists, none of whom holds this or an analogous view, and takes up English-language idealists like Berkeley, and Bradley (Rockmore 2001, 345). So unlike the idealists that Moore has in mind, Hegel does not explicitly discuss the question of an external world.[9] Instead, he replaces the concept of a noumenon with

9 Rockmore gives three clues on the position of Hegel as acknowledging something outside of the mind: "To begin with, in the so-called Differenzschrift, his first philosophical publication, he advances the view, which he never later renounces, that philosophy begins in (ontological) difference. His position can be simply characterized through the complex assertion that epistemological unity depends on ontological disunity, or difference. In this connection, it is crucial to note the distinction, on which Hegel insists, between the opposition between subject and object, in which the latter is simply defined as negation of subjectivity, and claims to know mind-independent objects as they are, which Hegel simply rejects. Second, in the *Phenomenology of Spirit* Hegel presents a phenomenology of the experience of consciousness. The stage-wise development of consciousness includes consciousness, self-consciousness, and reason, whose final phase is absolute knowing (absolutes Wissen), in which the subject is said to know and to know that it knows its relation to what is other than itself. Sense-certainty, the initial phase of consciousness, tells us that something is, or exists, hence acknowledges the existence of the external world but not what it is, which further requires perception. Third, the same basic position is further developed in the *Encyclopedia of the Philosophical Sciences* where Hegel provides an important account of the 'Positions of Thought to Objectivity.' It is incorrect to tax Hegel with denying the existence of the external world in any way, for instance, by reducing it to spirit. Moore thinks that either one denies the existence of the external world, or one must acknowledge its existence and be able to

phenomenology. By this, I mean that Hegel is primarily interested in understanding how an object appears to the subject as an object in itself. However, the 'in-itself' is something that the subject constructs or produces. In Hegel's framework, the noumenon, as a presumed real and external entity, becomes part of the phenomenon. It belongs to the consciousness that comes to realize that a sensed, perceived, or conceptualized object is always dependent on consciousness itself.

On the other hand, Moore's interpretation of external reality leads him to the position that "sensation is always sensation of something," implying that we are always in immediate contact with external things (Rockmore 2001, 345). However, the problem, as Hegel sees it, is not merely the question of the existence of the external world, as Moore suggests. Instead, Hegel recognizes that our understanding of this outside world may be fundamentally distorted. The challenge to be addressed is how we can establish a reliable connection between these distorted sensations, perceptions, or conceptions of objects and a mind-independent object. According to Hegel, we are currently unable to establish such a reliable connection. Nevertheless, it is a Hegelian viewpoint that natural consciousness perceives objects as immediate, but this immediacy is subjective to us. As Hegel demonstrates in the introduction to the *Phenomenology of Spirit*, the division between the 'in-itself' and the 'for-itself' is a division that exists within consciousness.

This significant observation serves as a motivation for Hegel's project of exploring self-consciousness and its manifestations in historically evolved societies. Rockmore argues that Moore's oversimplification of Hegel's idealism, favoring immediate contact with reality, had an influence on later debates within analytic philosophy (Rockmore 2001, 345). Historically, Moore's perspective has shaped the discourse that ultimately led to Sellars' intervention with his theory of the 'myth of the given'.

1.3 *The Neo-pragmatic Turn and Sellars' Linguistic Rationalism*

Sellars' philosophical framework marks a departure from the representationalism of Russell and Moore. His philosophy, known as linguistic rationalism, can be seen as an initial return to Hegelian thought within the analytic debate. However, it is important to note that Sellars' project diverges from Hegelian

know it as it is. The third possibility, which he does not consider but which Hegel defends, is that the existence of the external world can be conceded, or at least not denied, although no claim is made to know it as it is. There is an enormous difference between the traditional realist claim to know the mind-independent external world and the weaker" (Rockmore 2001, 350).

idealism. Sellars' linguistic rationalism aligns with a form of neo-pragmatic nominalism, which relativizes the influence of reality and emphasizes the role of humans and their social interactions in discourses. While Sellars' viewpoint resonates with Hegelian notions of the social, it does not fully connect with the broader idealist framework and historicist ideas that Hegel presents. Furthermore, Sellars' own account of linguistic rationalism, as extensively discussed in scholarly literature, may be incompatible with his conception of picturing, which aims to establish a relationship with a mind-independent reality and thereby challenges idealism. Picturing reflects Sellars' realism and scientism, contrasting with his nominalism. Both elements, nominalism and picturing, are essential for comprehending Brandom's conceptual realism as well. In fact, Brandom also seeks to merge neo-pragmatic, nominalist ideas with an approach to reality through his conceptual realism. However, as Brandom further develops his theory of inferentialism, he departs from Sellars' notion of picturing. Consequently, my focus will be on Sellars' linguistic rationalism as a response to the analytic tradition initiated by Russell and Moore, and his account of picturing, which arguably contradicts his overall efforts. Prior to delving into Sellars' ideas, I will provide a brief description of the historical transition from Moore to neo-pragmatism.

Moore's stance on the issue of immediacies is closely tied to his assertion of the mind-external world, which contrasts with the more pragmatic views put forth by the later Wittgenstein. According to Wittgenstein, it is crucial to question whether our "claims to know are mediated by a frame of reference" (Rockmore 2001, 346). In addition to highlighting the impact of the social dimension on our act of making knowledge claims, Wittgenstein suggests that the pseudo-problem of a mind-external world should not be a concern for philosophers. In our everyday practical lives, we rarely make explicit assertions about a world that exists independently of our perception. Instead, our discussions revolve around how we experience the world and how we can negotiate and share these perspectives with others. While Moore certainly influenced the analytic tradition, Wittgenstein's objections to Moore's ideas played a significant role in fostering a broader rejection of ideal language philosophies and their reliance on a type of raw empiricism. This rejection gave rise to the holistic movement, which represented a shift towards considering the contextual conditions and social aspects that shape our utterances and language use.

Rockmore distinguishes three forms of contextualist pragmatism: "Quine is concerned with how we can reasonably be said to know; Rorty, who is an epistemological skeptic, is concerned with why we can reasonably be said not to know; and Putnam is occupied with what and how we know" (2001, 356). Though all of these contextualist pragmatists develop important arguments against analytic philosophy's representationalism, it has been argued that they are "clumsily lumped together as neopragmatists" (Long 2002, 46).[10] The term 'neo-pragmatism' encompasses a diverse range of thinkers, forming an "unlikely collection of bedfellows" (Rockmore 1993, 277). This broad spectrum also includes Sellars, who is arguably the most influential figure for Brandom. However, it remains uncertain whether Sellars, like the other philosophers, can be fully categorized as a neo-pragmatist. On one hand, Sellars aligns with Wittgenstein's contextualist notion that a non-ideal language serves as the primary frame of reference. He rejects the raw empiricism found in early analytic philosophy and introduces linguistic coherentism. However, Sellars' linguistic coherentism encounters the problem of "frictionless spinning in a void" (McDowell 1996, 50).[11] Sellars, therefore, introduces a parallelism between mind and world through his concept of picturing. This parallelism keeps appearances, which reside within language, distinct from truth, which is the object of reference. Consequently, Sellars violates a crucial criterion of pragmatism (Long 2002, 46).[12] So far as we are concerned, this means that Sellars still holds on to the question of an ahistorical, objective criterion for knowledge that is independent of our actions.

Sellars stands out as one of the early figures who breaks away from the lineage of early analytic philosophy through his influential concept known as 'the Myth of the Given'. This concept fundamentally challenges representationalism as a metaphysical ideology that erroneously assumes the existence of a given at the level of sensory experience. Sellars contends that senses alone cannot provide the necessary meaning to our semantic concepts. In doing so, he tackles the historical problem of a rigid dualism between the formal side

10 Long, Joseph W. "Who's a Pragmatist: Distinguishing Epistemic Pragmatism and Contextualism." *The Journal of Speculative Philosophy* 16 (1), 2002: 39–49. http://www.jstor.org/stable/25670401.

11 McDowell, John. *Mind and World*. Cambridge: Harvard University Press, 1996.

12 The comparison to the original pragmatism is only reasonable to some degree, for the original pragmatist movement started with deflationism. According to Williams, "the pragmatists associate representationalism with an excessive intellectualism that tends to divorce knowledge from the activities of inquiry" (Williams, Michael. "Pragmatism, Sellars and Truth." *Sellars and his Legacy*. Edited by James R. O'Shea. Oxford: Oxford University Press, 2016, 223). Instead of assuming that theoretical claims are the foundation of knowledge, the classical pragmatists focus on truth as action (see Williams 2016, 223). Consequently, 'making true' is an activity in which we show that something helps us in coping with practical problems (see Williams 2016, 223). But 'making true' is only one specific field of interaction. There is therefore a problem to call the contextualist, pragmatic movement pragmatism since they either maintain a theoretical approach to truth or fall entirely into skepticism.

of knowledge and its supposedly independent material information. Sellars raises doubts about the complete independence of sensory receptivity from the conceptual framework that shapes our understanding. In light of this, he explores the normative status of knowledge and its implications: "characterizing an episode or a state as that of knowing, we are not giving an empirical description of that episode or state; we are placing it in the logical space of reasons, of justifying and being able to justify what one says" (Sellars 1997, 76).[13] Sellars offers an alternative perspective by reframing reference not as an isolated and unrelated given, like an empiricist would, but as an integral part of a broader network of reasons. For Sellars, any empirical episode that contributes to knowledge is situated within this network of reasons, making it inherently "a linguistic affair" (Sellars 1997, 63). In this view, knowledge becomes deeply intertwined with social practices within the space of reasons. It is through this space of reasons that we are able to communicate and share empirical episodes, employing what can be referred to as material inferences that arise from the exchange of reasons (see Rorty 1997a, 4).

In his introduction to Sellars' *Empiricism and the Philosophy of Mind*, Rorty writes about Sellars' impact: "Sellars's attack on the Myth of the Given was a decisive move in turning analytic philosophy away from the foundationalist motives of the logical empiricists" (Rorty 1997a, 5). Moreover, by questioning logical empiricism, Sellars' objection also represents a turn towards Hegel.[14] Sellars draws upon the holistic arguments put forth by Hegel in the chapter on sense-certainty in the *Phenomenology of Spirit*. Hegel's contention was that the notion of direct knowledge, or what he referred to as 'the immediate,' is a fallacy. This fallacy was perpetuated by a representationalist framework that ultimately failed to provide a justification for its position (see Rockmore 2001, 360). In a similar vein, Sellars introduces the concept of semantic assertibility, which aims to establish truth grounded in our normative commitment to rules. In *Science and Metaphysics*, Sellars states his central thought:

> for a proposition to be true is for it to be assertible, where this means not *capable* of being asserted (which it must be to be a proposition at all) but *correctly* assertible; assertible, that is, in accordance with the relevant

13 Sellars, Wilfrid. *Empiricism and a Philosophy of Mind*. Cambridge: Harvard University Press, 1997.

14 Rockmore writes on Sellars' rejection of empiricism: "His [Sellars'] central point seems to be that we do not and cannot depend, as classical empiricists believe, for the justification of claims to know on immediate experience whose probative force is undermined by the rejection of the myth of the given" (2001, 348).

> semantical rules, and on the basis of such additional, though unspecified, information as these rules may require [...]. 'True', then, means *semantically* assertible ('S-assertible') and the varieties of truth correspond to the relevant varieties of semantical rule.
>
> SELLARS 1968 §26, 101[15]

Semantic assertibility implies that truth is contingent upon the semantic rules that govern a given framework. Truth is not solely determined by the correspondence between words and facts, but rather it is an action rooted in our adherence to these semantic rules. By establishing sets of semantic rules, we can define the relationships between different predicates. For instance, the commonly used phrase 'if it rains, the street becomes wet' suggests a specific connection between the predicates 'rain' and 'wet'. Consequently, we can infer that if 'wet' is the opposite of 'dry', then 'rain' cannot be 'dry'. Through this process, we derive a network of semantic relations without relying solely on sense data as our foundation.

Kolb highlights how the concept of material inference allows for both formal and meaningful connections within our semantic network. Different predicates possess distinct formal relationships with other members of their genus. For instance, 'red' and 'green' are related as species of 'color', while 'loud' and 'piercing' are species of 'noise'. Additionally, 'noise' and 'color' are species of 'sensations', and so on (Kolb 1978, 386).[16] So predicates receive meaning by virtue of material inferences. They build, for example, classes of predicates, since their relations would suggest contradictions otherwise.

Overall, Sellars' point is that we interpret experiences coherently with regard to our semantic net, but we do not simply represent a mind-external reality. There is no inference back from our semantic assertions to a mind-external reality. Though there is a relation between names and the world, names are only "the basic elements which picture particular entities within the world" (Kolb 1978, 388). Beyond the aspect of picturing, they play different roles in a "network of linguistic types" (Kolb 1978, 388). In other words, the semantic net, which is defined by the relations of the terms, will determine the meaning of the used names (see Kolb 1978, 388). The meaning of a name does not represent a single fact outside of our semantic net but becomes involved in various

15 Sellars, Wilfrid. *Science and Metaphysics—Variations on Kantian Themes*. New York: Humanities Press, 1968.

16 Kolb, David. "Sellars and the Measure of All Things." *Philosophical Studies: An International Journal for Philosophy in the Analytic Tradition* 34 (4), 1978: 381–400. http://www.jstor.org/stable/4319263.

structures that help us with our practical orientation. We can identify this as Sellars' linguistic rationalism or coherentism (Kolb 1978, 388).

The major problem of semantic assertibility, however, is that it "provides no way to adjudicate whether the semantical rules of one linguistic framework are more adequate than the rules of another" (Levine 2007, 248).[17] In other words, Sellars' position implies that multiple semantic nets can exist, each internally coherent within itself, but potentially incompatible with other semantic nets. This perspective raises a significant concern known as the problem of frictionless coherentism, as also pointed out by Rockmore:

> The space of reasons relies on coherence, or the interrelation of concepts in a conceptual framework. Yet since the coherence in question cannot rely in any way at all for justification on the given [...], it is an instance of what McDowell, who is otherwise sympathetic to Sellars, calls 'unconstrained coherentism' [... A] theory can be coherent but false. Many individuals in mental institutions have coherent worldviews.
>
> ROCKMORE 2012[18]

Rockmore implies that the assessment of one framework being 'more true' than another would be contingent upon the semantic rules of that particular framework. Sellars himself acknowledges the issue of "linguistic idealism" and raises the question: "If facts belong to the linguistic order and the world consists of facts, must not the world belong to the linguistic order? And is not this a miserable absurdity?" (1962, 41).[19] This aspect highlights a limitation in his linguistic system, where our access to mind-external objects beyond language is constrained. This deficiency, reminiscent of idealism, prompts Sellars to introduce the concept of 'picturing' (see Levine 2007, 248) as a means to restore a connection to a mind-external reality (see Williams 2016, 224). This, however, raises the question of whether Sellars deviates from his linguistic rationalism.

Rorty expresses his dissatisfaction with Sellars' attempt to introduce the concept of picturing, stating, "many of us whose minds were formed by reading Sellars think of Sellars' doctrine of picturing as an unfortunate slide back into

17 Levine, Steven. "The Place of Picturing in Sellars' Synoptic Vision." *The Philosophical Forum* 38 (4), 2007: 247–269. Accessed November 23, 2017. http://onlinelibrary.wiley.com/doi/10.1111/j.1467-9191.2007.00266.x/full. I used the draft provided by Levine from Academia.edu https://www.academia.edu/3304708/Sellars_on_Picturing.

18 Rockmore, Tom. "The Pittsburgh School, The Given and Knowledge." *Social Epistemology Review and Reply Collective* 2 (2), 2012: 29–38. http://wp.me/p1Bfgo-Ak.

19 Sellars, Wilfrid. "Truth and 'Correspondence'." *The Journal of Philosophy* 59 (2), 1962: 29–56. doi:10.2307/2023294.

representationalism" (Rorty 1988, 216).[20] Despite a series of attempts to defend Sellars' account of picturing (Levine 2007, Sicha 2014),[21] 'picturing' remains controversial (see DeVries 2016),[22] especially since it also raises the question of how 'picturing' is related to Sellars' equally controversial notion of scientism.[23] After decades of debate, even the disciples of Sellars are divided on it.

Given Brandom's rejection of Sellars' account of picturing, it is more important to concentrate on the aspects that influence Brandom's conceptual realism. Both Brandom and Sellars share a common objective: to challenge 'bad' idealism, or, as Sellars writes about it: "Though full of important insights, Idealism is, therefore, radically false" (Sellars 1968, 143).[24] Basically, Sellars argues that we cannot form a concept of reality solely on the basis of our mental expressions. The sciences and their understanding of reality through causality, however, are superior to a merely mentalist approach that only experiences causal connections.[25] However, it is important to clarify that Sellars views the sciences as offering a more complete account of reality, incorporating both mental and physical dimensions. In his autobiographical notes, Sellars writes on this role of science "the solution of the puzzle lay in correctly locating the conceptual order in the causal order and correctly interpreting the causality involved" (Sellars 1975).[26] Unlike Hegel, Sellars does not seek to explain sciences through the stages of natural consciousness but relies on the concept of causality that is tied to his scientific worldview. DeVries, like many other interpreters, critiques Sellars' anti-idealist argument for simply dismissing the manifest image of the mind:

20 Rorty, Richard. "Representation, Social Practise, and Truth" Philosophical Studies: *An International Journal for Philosophy in the Analytic Tradition* 54 (2), 1988: 215–228. http://www.jstor.org/stable/4319982.

21 Sicha, Jeffrey F. "Defending the Unpopular Sellars: Picturing and 'The Descriptive'." *Journal of Philosophical Research* 39, 2014: 127–163. DOI 10.5840/jpr201472914.

22 DeVries, Willem A. "Wilfrid Sellars." *The Stanford Encyclopedia of Philosophy.* Edited by Edward N. Zalta. 2016. https://plato.stanford.edu/archives/win2016/entries/sellars/.

23 For the relation between 'picturing' and Sellars' scientism, see, for example, Sicha (2014, 128).

24 DeVries writes about this quote: "the notion of picturing, Sellars believes, enables him to reject idealism once and for all" (DeVries, Willem A.: "Getting Beyond Idealisms." *Empiricism, Perceptual Knowledge, Normativity, and Realism.* Edited by Willem A. DeVries. Oxford: Oxford University Press 2009: 211–247, 242).

25 I interpret Sellars as attributing causality to the scientific worldview while ascribing causal connections to the manifest image.

26 Sellars, Wilfrid. "Autobiographical Reflections." *Action, Knowledge, and Reality: Critical Studies in Honor of Wilfrid Sellars.* Edited by Hector-Neri Castañeda. Indianapolis: The Bobbs-Merrill Company, 1975 http://www.ditext.com/sellars/ar.html.

> He needs a positive argument to establish that the manifest image cannot develop such an explanation and that we must turn to a radically novel scientific framework to do so. The manifest image has shown itself in the past to be a flexible tool for coping with reality, capable of growth and development in order to accommodate an ever richer understanding of the structure of reality and our relation to it.
>
> DEVRIES 2009, 242

Sellars' approach to bridging the gap between the internal, manifest image of linguistic affairs and the scientific image remains unclear. DeVries criticizes Sellars' notion of the "really real" for becoming "unrecognizable" when he introduces a distinct division between the manifest and scientific images (DeVries 2009, 242). On the one hand, Sellars acknowledges the limitations of our perspectives within the manifest image of our intra-linguistic experience. On the other hand, he argues for the importance of scientific understanding, which allows us to evaluate our linguistic actions externally. However, the question arises: How can we achieve this external evaluation when our scientific communication still relies on intra-linguistic experience?

To understand Brandom's motivations for establishing objectivity, it is essential to examine the relationship between picturing and Sellars' scientism, as well as to identify the challenges he faces in his pursuit to overcome idealism. 'Picturing' is introduced as an outside view on our language. For this, Sellars remarks that "the hook-up of a system of rule-regulated symbols with the world is not itself a rule-governed fact, but [...] a role in the rule-governed linguistic system, and [...] a role in the structure of tied sign responses to environmental stimuli" (Sellars 1949, 310).[27] Though Sellars argues that the normative constitution of our language is not simply caused by the world, he assumes here that something justifies the linguistic system from the outside. In this sense, it is a 'response' to the environment. Following these claims, we have to see that our "mental episodes" have to be understood with regard to "language–entry transitions" (Levine 2007, 250). So, for example, children learn to interpret the noise 'blue' similar to dogs learning the noise 'bone' (see Sellars 1949). Then, at some point, the noise 'blue' fulfills, as Sellars puts it, a "function for us in a system of rule-regulated symbol activity, and it is a *word*, a linguistic fact, a rule-regulated symbol" (Sellars 1949). In other words, the stimulus becomes a word. But before it becomes a word, it is a stimulus. This means

27 Sellars, Wilfrid. "Language, Rules and Behavior." *John Dewey: Philosopher of Science and Freedom*. Edited by Sidney Hook. New York: The Dial Press, 1949: 289–315. http://www.ditext.com/sellars/lrb.html.

that though the signs of our language signify something within the system of language, they also picture an outside world by virtue of "causal uniformities" (see Levine 2007, 254). Levine underlines this difference between words and causes: "The causal hook up between language and world that takes place through our being inculcated into certain non-conceptual uniformities is one of *picturing* while the rule-regulated aspect of a language is one of *signifying*" (Levine 2007, 254). 'Signification' is, according to Levine, about the relation between the items of our intentional action (see Levine 2007, 254), while 'picturing' is about the relation between language items as items within a causal nature.

Sellars' original point was that representationalist truth-paradigms had been neglecting the aspect of signification. Words were thought to receive meaning by virtue of representation. As a consequence, representationalist theories defend a conception of truth that is independent of the action of the speaker and the normative ways of conceptualizing. With his linguistic rationalism, Sellars offered an alternative construction of meaning from within language. Picturing, however, is another story. According to Sellars, picturing does not interfere with signification and the normative side of our language. Instead, picturing reconstructs language as an "*isomorphic* relation between two types of *objects*" (Levine 2007, 254). These two objects are natural linguistic words and natural entities. Sellars shifts towards a naturalist view on language in "Truth and 'Correspondence'":

> If picturing is to be a relation between objects in the natural order, this means that the linguistic objects in question must belong to the natural order. And this means that we must be considering them in terms of empirical properties and matter-of-factual relations, though these may, indeed must, be very complex, involving all kinds of constant conjunctions or uniformities pertaining to the language user and his environment.
>
> SELLARS 1962, 44[28]

Here, we abstract from the normative, internal structure of language and look at linguistic episodes as objects in the natural order. According to Sellars, it is now about the 'natural-linguistic object' (see Sellars 1962, 44).[29] As Sellars

28 Sellars, Wilfrid. "Truth and 'Correspondence'." *The Journal of Philosophy* 59 (2), 1962: 29–56.

29 Of course, Sellars discusses further problems that occur with regard to such a changed perspective: "But what if, instead of construing 'picturing' as a relationship between facts, we construe it as a relationship between linguistic and nonlinguistic objects? The very formulation brings a sense of relief, for in everyday life we speak of pictures of things or persons, not of facts. Roughly, one object or group of objects is a picture of another

specifies "linguistic episodes have not only logical powers" (Sellars 1962, 44). They also have "matter-of-factual characteristics, e.g., shape, size, color, internal structure" and "exhibit empirical uniformities both among themselves and in relation to the environment" (Sellars 1962, 44). Thus, they are "objects in nature with respect to their matter-of-factual characteristics" (Sellars 1974, 52, see also Levine 2007, 254).[30] We have to keep in mind that the one perspective of language is that it consists of natural objects. This perspective is different from the normative view of language which focuses on our normative commitments and the construction of linguistic objects.

With Sellars' approach to signification and picturing, we have to deal with two perspectives: "the rule-governed assertibility of semantically significant items" and the "natural objects in the real order" (Levine 2007, 254). Epistemologically, the two perspectives on language are supposed to escape the problem of ontological difference, i.e., the connection between matter and form. On the one hand, we explain language from the inside. On the other hand, we look at it as a natural object. The important question, however, arises: how can we think both perspectives on language together? How is the "rule-regulated calculus, and a cluster of conditioned responses which binds us to our environment" related to our normative activity of conceptualizing (Sellars 1949)? Kolb, for example, argues that Sellars must assume that sciences will describe everything in terms of atomic phrases. Sellars does support such interpretations by different statements that are not entirely clear in their dialectical purpose. For example, he notes that with his concept of picturing, we obtain an "Archimedean point outside the series of actual and possible beliefs in terms of which to define the ideal or limit to which members of this series might approximate" (Sellars 1968, 142). The Archimedean point has to be understood in terms of the organism's success by using its language. Kolb reconstructs the relation in the sense that the "complex isomorphism between the particulars in the world [...] and the atomic sentences in the world-story" (Kolb 1978, 393) can be understood if we see that "the linguistic array is used by an organism to guide its behavior" (Kolb 1978, 393). Then, the linguistic sphere is motivated by our success in the natural world to which we approximate. A successful linguistic action occurs only if a sentence can resemble the

object or group of objects. Yet since objects can picture objects only by virtue of facts about them (i.e., only by virtue of having qualities and standing in relations), it may seem a quibble to insist that it is objects and not facts that stand in the picturing relation. It is not, however, a quibble, but the heart of the matter" (Sellars 1962, 44).

30 Sellars, Wilfrid. 1974. "Some Remarks on Kant's Theory of Experience." Essays in Philosophy and Its History. Dordrecht-Holland: D. Reidel Publishing Company.

same structure of the external natural world by picturing. Does Sellars say here that our linguistic apparatus must be explained with regard to mind-external reality?

Sellars' account of picturing is, indeed, grounded on a naturalist presupposition, as pointed out by DeVries, who questions whether an "empirically meaningful language game must contain a subgame, the rules of which effectively require the construction of a map, a representation of the world related to the world in naturalistically respectable ways" (DeVries 2005, 55).[31] While Sellars may not completely naturalize language, there are several criticisms that need to be considered. Firstly, although organisms may use language to guide their behavior at times, it does not necessarily imply that we can causally explain the internal structure of language solely based on this behavior. The conditions for entering into language games could be arbitrary and not essential for the development of internal language moves. Secondly, the constitution of language cannot be solely explained as a tool for guiding behavior, as language is only successful in guiding behavior in specific instances. It is possible that language originated with an aesthetic or social function, such as supporting cooperation among people or facilitating discussions about meaning structures independent of the mind-independent world. The instrumentalist view of language, which claims that its main purpose is to achieve success in the external world, neglects these alternative interpretations of language. Thirdly, language could be an additional function of the brain that evolved as an unnecessary appendage, unrelated to the external world. In this case, language would lack intrinsic function and instead serve as a tool utilized by an underlying will for a specific purpose. If these arguments hold merit, the naturalist argument that reduces language to a practical representation of physical reality is just one hypothesis among many. While this hypothesis is relevant for biologists seeking to explain language, we must first clarify how we use language for success in the natural world before accepting such a naturalist assumption.

However, even if we concede that language aids us in guiding our behavior only in specific instances, it raises the question of whether we truly comprehend physical reality as it is or simply improve our ability to grasp it. Sellars illustrates this idea with the example of a robot that scans its environment using radiation. As the robot moves and gathers data, it accumulates more information about its environment and gains a better understanding of how the world

31 DeVries, Willem A. *Wilfrid Sellars*. Chesham: Acumen Publishing, 2005.

operates (Sellars 2007, 220).[32] Kolb writes about this particular formalism of a machinist conception of language: "Once we give up the notion of an intentional relation between the mind and the world, the isomorphism of picturing forms the link between the world and our brains as organic computers juggling symbols in accordance with formal rules" (Kolb 1978, 393). The concept here is that we engage in the process of mapping the external world, and although the reconstruction is influenced by our semantic rules, we progressively develop a more precise representation of reality. For instance, embarking on a journey from Chicago to San Francisco would not yield a purely contingent depiction (see Kolb 1978, 393). However, unexpected occurrences such as "Kansas City or the Rocky Mountains [...] small deviation in the orbit of Mercury" (Kolb 1978, 393) would expose inaccuracies in our maps. By making these corrections, our representation becomes more nuanced (see Kolb 1978, 393) and consequently more accurate. In other words, from a naturalist perspective, language is not solely a relativistic framework. According to Sellars' viewpoint, it is also a naturalistic entity with an objective purpose.

Nevertheless, the notion that mere data collection leads to a deeper understanding of a mind-independent world is flawed. It brings to mind Putnam's exploration of the 'brains in a vat' scenario and the question of whether all our knowledge claims about an external reality are inherently flawed. Instead, we can only gather information about the internal world that is presented to us, leaving us uncertain about how our internal world relates to an external reality. To illustrate this further, let us consider another example that resembles Sellars' robot scanning its environment to gain knowledge.

Imagine a person tasked with solving a labyrinth on a computer screen. The screen does not show any lines representing the walls; only a cursor indicates the person's position. This situation is akin to our mind's internal reality, where we only possess information gathered within the labyrinth. One might argue that we gather information about the labyrinth by moving around. However, this view is incorrect because our gathered knowledge can be entirely false. The solution to the labyrinth may not depend on our actions. For instance, let us assume that after a random amount of time, the labyrinth will be solved regardless of what the person does. After x minutes, the screen will indicate the person's success. Consequently, all the cursor movements made by the person become inconsequential for solving the labyrinth. Yet, they would have collected arbitrary data about how to solve the labyrinth and might even

32 Sellars, Wilfrid. "Being and Being Known." *The Space of Reasons, Selected Essays of Wilfrid Sellars.* Edited by Kevin Scharp and Robert Brandom. Cambridge: Harvard University Press, 2007, 220.

present it as knowledge. In fact, however, there is no solution. For example, person A could claim to person B, who also attempted to solve the labyrinth, "I succeeded by moving up two squares and then turning right." Both of their knowledge systems would be false in terms of solving the labyrinth, resulting in the creation of pseudo-knowledge.

The crux of the matter is that without external knowledge about the labyrinth, these individuals cannot determine the extent of their errors. While we could argue that they produce coherent systems, there is no direct connection between coherence and knowledge about an external reality. If we were indeed brains in a vat, our entire system of current knowledge could collapse. We might be in a labyrinth, rendering it absurd to assume that our exploration necessarily unveils reality as it truly is. Similarly, claiming with certainty that we are brains in a vat, living in a simulation, or trapped in a labyrinth would be absurd. What truly concerns us is the empirical reality we experience.

There are further factors that raise concerns about Sellars' ideas. Firstly, the example of a robot scanning its environment relies on our assumptions about normal physical extensions and our typical movement through time in a specific direction. However, these presuppositions about a mind-independent reality lack a solid foundation. In fact, concepts such as extension and movement are intertwined with our understanding of ourselves as physical bodies. While it is challenging to determine whether we are solely physical bodies, Sellars' theory of neurological states implies that we can explain the mind in relation to a physical, mind-independent world. Sellars confidently asserts that through neurophysiological states, we will come to understand what it means to be a thinking entity. He states, "I submit that as belonging to the real order it is the central nervous system, and that recent cybernetic theory throws light on the way in which cerebral patterns and dispositions picture the world" (Sellars 2007, 228). This suggests that, eventually, we will be able to translate the structures of our mind into a scientific theory that comprehends the true nature of the mind. I think this is very controversial. Secondly, we must consider that the physical world may be subject to change. Therefore, any statement we make about reality may hold true only for a specific moment in time. Thirdly, we must acknowledge that reality is vast and encompasses a multitude of potential events. It is irrational to assume that by grasping only a portion of it, we automatically gain a comprehensive understanding of the entire reality and subsequently improve our knowledge of it.

While we have collectively agreed upon certain rules that have proven successful in the aspects of experience we have encountered thus far, we still lack a secure understanding of what reality truly is. Instead, we rely heavily on our individual and collective experiences to create limited knowledge. As a result, our achievements thus far have primarily led to the establishment of a relatively stable society based on the cooperation and shared experiences of its members. To advance in the pursuit of knowledge, it might prove challenging to fully unravel the complexities of reality. Before we can dedicate ourselves to this task, it becomes crucial to attain a deeper comprehension of our own identity and nature to establish at least relative knowledge with regard to our experiences successfully.

To illustrate the interdependence between knowledge and the intrinsic condition of being human, let us revisit Kolb's example of a journey from Chicago to San Francisco. By considering historicist and anthropological aspects, we can better navigate knowledge claims. Traveling from Chicago to San Francisco entails a clear goal, which is reaching a specific physical location. It involves traversing a well-defined space-time continuum with a limited number of foreseeable events that might hinder the journey. The story of acquiring a progressively more accurate understanding of this particular trip is relatively straightforward. However, the human endeavor and our voyage through space-time, history, evolution, and beyond offer far more than simply arriving in San Francisco. While it can be argued that we do create knowledge, the process of attaining better knowledge is not merely a result of haphazardly collecting data during an arbitrary trip through physical space. Not only may our data collection be flawed due to our uncertainty about the true nature of the conditions under which we gather information, but the creation of knowledge is also contingent upon the goals we define as human beings. Consequently, the pursuit of knowledge hinges, as I will later illustrate with Hegel, on how we position ourselves in relation to knowledge itself. Knowledge is not primarily reliant on external objectives but rather on the introspective question of our identity as human beings within our historical context. It is in our roles as human beings that the events occurring during a journey from Chicago to San Francisco become significant. Ultimately, any assumption regarding a particular type of data can only furnish us with a clearer understanding of what reality means to us. This understanding aligns well with the perspective I will introduce in Chapter Three on Hegel. In contrast, Sellars focuses primarily on exploring the reality external to the mind. Although his viewpoint is more nuanced, the resolutions he presents revolve around the tension between his linguistic nominalism and his belief in comprehending a mind-independent reality. In the subsequent paragraphs, I will discuss this conflict.

Given my objections, one could argue that Sellars assumes picturing only for a small amount of language entry conditions, while he believes that the normative commitments between humans do not always relate to these conditions. If this interpretation is true, then Sellars' account of picturing is an amendment. It would not change the core of his linguistic coherentism. However, I argue that Sellars does not want to restrict his view to the language-entry-conditions. Kolb, for example, makes Sellars' commitment to the original atomism of analytic philosophy clear: "while any partial or complete world-story would be a system of juggled symbols, the true world-story will be the one whose manipulations end in the pattern of atomic sentences which is isomorphic with our world" (Kolb 1978, 393). In other words, the ultimate picture is then the unification of "everything under one set of laws and name[s]" (Kolb 1978, 393). The world is physical and consists of only one set of laws. It might be a transcendental necessity for sciences to assume a possible unification of our world-views, but I think that Sellars, if he follows such a strict scientism, conflates the necessity of a unity with the real possibility of achieving this in a final theory. He would take a possibly necessary regulative idea for sciences to be reality. According to Kolb, Sellars' "full world-story" identifies all particulars and utilizes all possible predicates (see Kolb 1978, 390). It means that within the manifest image, we are speaking within a minor variation, a dialect, of what is real truth-talk (see Kolb 1978, 390). Sellars' view implies then that we always come closer to truth, similar to adding more and more details to our picture of reality. Truth is then physical reality and the manifest image of the world must be embedded in this physical reality.

Though such an ideal can be guiding for discourses and might be necessary—since we need a criterion for truth discourses in, let us say, sciences—this does not mean that such an ideal is correct and independent of history. Under these conditions, Sellars would transform the need for a criterion for practical endeavors, and our choice of some criteria in historical contexts into the existence of an absolute criterion, in this case, to tell the story of physical time and space. Oddly enough, Sellars also writes against such kind of physical reductionism:

> then one would expect there to be a sense in which the scientific picture of the world *replaces* the common-sense picture; a sense in which the scientific account of 'what there is' *supersedes* the descriptive ontology of everyday life [...] But, *speaking as a philosopher*, I am quite prepared to say that the common-sense world of physical objects in Space and Time is unreal—that is, that there are no such things.
>
> SELLARS 1997, 82–83

Sicha in attempting to defend "his Sellars" claims that "the scientific image will [not] replace the entirety of the manifest image" (Sicha 2014, 128). He remarks that also for Sellars the world itself includes minds, while only his descriptive ontology avoids the idea of the mind. Indeed, Sellars seems to limit science to the empirical field of explanation: "in the dimension of describing and explaining the world, science is the measure of all things, of what is that it is, and of what is not that it is not" (1997, 83). As Sicha emphasizes, the explanation might only apply to physical objects and not to normative statuses (see Sicha 2014, 130).

Yet, all these relativizing aspects of Sellars' physical worldview cannot deny Sellars' overall restrictive view of science, as expressed in the "descriptions of spatio-temporal entities (or events) of a very simple sort" (Kolb 1978, 382). This view will always stand in contradiction to Sellars' linguistic coherentism. Kolb makes this contradiction very clear. He argues that Sellars' "skeletal relational predicate [...] which signifies the fundamental type of order in which the individuals to which the language can refer must stand" (Kolb 1978, 374) requires mind-external objects. Sellars' linguistic coherentism, however, denies such mind-external objects, at the same time.

Kolb reconstructs Sellars' major argument for a mind-external reality: "Since material rules of inference make world-stories possible, there could be no world-story whose individuals do not stand in an appropriate skeletal relation" (Kolb 1978, 374). So all individuals stand in material relations to each other, since material conditions are the precondition of our semantic skeleton. The assumption is that if these material conditions remain unfulfilled, then there would be no meaning at all. The argument is then: since we have meaning there must be a reality that fulfills these material conditions (see Kolb 1978, 395).

The argument, as Kolb discusses further, does not work. The problem is that a mere reference to mind-independent individuals is not possible (see Kolb 1978, 395). It is a misapplication of our normative intentionality to assume individuals in the mind-external world. In other words, individuals are conditions for communication, but this does not mean that the mind-external world consists of individuals. Concepts like 'individual' could only be abstract ways of coping with particular problems. To assume a reality of such individuals, and then to presuppose that our striving for a coherent picture refers ultimately to such simple entities is then, to use a Kantian term, a category fallacy. Interestingly, Sellars aligns with this counter-argument. Consequently, we must conclude that a nominalistic, normative system of semantic references and picture realism cannot be harmonized (see Kolb 1978, 396).

Here, it cannot be our task to mediate between right-wing and left-wing-Sellarsians. I mainly referred to Kolb as an early evaluation of Sellars' arguments. The question, however, remains of whether Sellars' linguistic coherentism and his account of picturing are simply incompatible. I believe I have raised some reasonable doubts regarding their compatibility. From a Hegelian perspective, as we will see, it is questionable to presuppose access to a mind-external reality. Sellars' attempt to defend mind-external realism is finally the reason that Rorty sees in Sellars' remainders of representationalist theories,[33] even though Sellars, according to Rorty, breaks radically with first-order-empiricism.[34] Rorty, who has been Brandom's dissertation director, goes even further than simply rejecting Sellars. He comes to the conclusion that afterall the failed attempts of the analytic school to solve the epistemological problems of representationalism, epistemology has to be dismissed overall. I argue that contrary to Rorty's epistemological skepticism,[35] Hegel offers a historicist perspective that relies on experience in order to give us knowledge. Epistemology might still be possible without representationalism. However, since Hegel is misunderstood, his historicist ideas play a minor role in this debate. Rather, what we have is Sellars' or even Brandom's 'Hegelian' philosophy that is still concerned with the problem of how to access mind-independent reality. As I claim, such philosophies tangle themselves in possibly unsolvable contradictions. Even though Brandom is less interested in Sellars' account of picturing, he claims to maintain a similar relation to mind-external reality. He will therefore, as I will argue, involve himself in similar contradictions.

According to Rorty, Sellars transitioned analytic philosophy from its Humean phase to its Kantian, but Brandom, Sellars' student, transitioned analytic philosophy from its Kantian to its Hegelian phase (Rorty 1997a, 3, 8, 9).[36] This image of transitioning is, however, questionable. On the basis of the development of analytic philosophy, we cannot, for example, say that the

33 Regarding the neo-pragmatist's break with the raw empiricism of a sensory given, Rorty famously distinguished two schools of analytic philosophy: the first school is interested into how we represent things as they are. The second school is interested in how we use language internally (see Rorty 1988, 215).

34 With regard to this empirical realism, Rockmore distinguishes three kinds: first, "direct knowledge of an independent object" (Rockmore 1997, 197), second, "indirect knowledge of an independent object through a dependent object" (Rockmore 1997, 197), and third, the denial of direct and indirect knowledge claims of a mind-independent world, but experience of mind-dependent objects. Rockmore argues that Hegel commits strongly to this "tertiary empiricism" (Rockmore 1997, 197).

35 I differentiate between skepticism and historicist relativism. While skepticism, according to my nominal definition, posits that no knowledge is possible and fundamentally denies any path to truth, historicist relativism, as previously discussed, allows for the possibility of a route to truth. It is important to note that in making this distinction, I do not discount the existence of meaningful versions of skepticism in the history of philosophy that may transcend this categorization.

36 Rorty, Richard. "'Introduction' to Wilfrid Sellars." In *Empiricism and the Philosophy of Mind.* Edited by Richard Rorty. Cambridge: Harvard University Press, 1997.

Kantian or Hegelian phase of analytic philosophy means that they resemble Kantian or Hegelian philosophy. It rather means that they employ themes of Continental philosophy in an analytic discourse that is interested in different problems motivated, as argued, by their rejection of what they assume to be idealism. By this, they invite problems that did not concern Hegel since he did not discuss access to a mind-external world. On the contrary, Hegel's idealism is grounded on the insight that we currently cannot access mind-external reality, which is, however, different from the claim that reality does not exist. As a consequence, Hegel's account cannot be easily transformed into a Neo-Hegelian mind-external realism of what is called the Pittsburgh school. Rather, as I argue, Hegel remains a strong competitor of them and Brandom.

1.4 *Rorty's Skepticism and Brandom*

As analytic philosophy is characterized by its lack of monolithic unity and is instead unified by a shared method and a focus on clarity, it is essential to acknowledge that certain branches within analytic philosophy challenge the notion of mind-external realism. This challenge, in turn, presents an opportunity to reconsider and reintegrate Hegelian perspectives into the discourse. In light of this paradigm shift, traditional views such as the belief that our concepts directly mirror nature lose their significance and become meaningless (see Rockmore 2001, 357). However, it is important to note that in my perspective, Rorty, as a critic of this view, erroneously interprets the associated arguments as leading to skepticism. Rockmore reconstructs his argument as follows: since knowledge is inherently representational and our means of investigating the source of these representations are confined to employing additional representations, we are ultimately unable to transcend the realm of representations themselves (see Rockmore 2001, 357). By rejecting conventional knowledge claims and embracing skepticism, Rorty believes that he aligns himself with the principles of Continental philosophy, which he now considers as a conversational philosophy: "They [*die Gesellschaft für analytische Philosophie in Deutschland*] correctly perceive that a thoroughgoing holism will sooner or later lead to a conversational view of philosophy" (Rorty 2003, 27).[37] The shift from the "representationalist view" towards a "contextualist theory of justification" is for Rorty a shift away from the problem of truth (Rockmore 2001, 357). Rockmore succinctly summarizes Rorty's position as follows:

37 Rorty, Richard. "Analytic and Conversational Philosophy." *A House Divided—Comparing Analytic and Continental Philosophy*. Edited by C. G. Prado. Amherst, NJ: Humanity Books, 2003, 27.

> If analytic foundationalism fails, then philosophy in all its many forms fails. Since philosophy is concerned with knowledge, and since analytic foundationalism fails, there is nothing interesting to say about knowledge.
>
> ROCKMORE 2001, 357

The argument implies that for Rorty analytic philosophy represents the final, most comprehensive, but failed attempt to deduce the foundations of our knowledge. However, it remains unclear why we should assume that we have exhausted all possible avenues in this pursuit, particularly considering that we have not continued exploring Hegel's historicist philosophy, which proposes the compatibility of relativism and the concept of truth. In my perspective, Rorty overlooks the possibility that circular epistemologies, such as Hegel's historical philosophy, could serve as an alternative to representationalism and skepticism. Therefore, I return to the notion that there exists a kind of knowledge that is relative to reflections on history—an idea rooted in Hegelian philosophy that Brandom, due to its analytic viewpoint, largely overlooks.

For similar reasons, it is questionable, as Rockmore points out, whether we can interpret the return to epistemic holism, as seen in Sellars, Putnam, and Brandom, as a return to Hegel. Rockmore distinguishes three forms of holism. First, "epistemic holism, or the Quinean view that theories meet their fate as a whole". Secondly, the "Wittgensteinian 'social' justification of claims to know in relation to a form of life". Lastly, "the Hegelian idea that claims to know are indexed to the historical moment" (Rockmore 2001, 363). Even though Rockmore sees similarities between Hegelianism and neo-pragmatism, he argues that basic differences are overlooked. It is similar to Hegel since pragmatist contextualism represents a denial of knowing the world directly as it really is (see Rockmore 2001, 365). Nevertheless, beyond this rejection of knowing a mind-independent reality, it does not develop the question of contextualism with regard to history but with regard to the social (see Rockmore 2001, 366).

The shortcomings of representationalism, which was once a foundational principle of analytic philosophy, led to the division of analytic philosophy into various programs: a holistic movement, a representationalist sect, and philosophers like Rorty who argue that philosophy is incapable of attaining any knowledge about knowledge itself (see Rockmore 2001, 357). Brandom agrees with Rorty's insights against representationalism, but, according to his attachment to Sellars, he does not commit entirely to Rorty's skepticism or to deflationism (see Rockmore 2001, 361). His book *Making It Explicit* directs us instead to social self-consciousness, which represents an attempt to overcome the conflict between content and formal, inferential semantics (see Rockmore 2001, 361). Bavaresco emphasizes this shift to the relationship of thought and action.

According to him, Brandom wants to explain semantic content in terms of the interactions between people:

> Brandom is associated with these strategies. He is interested in such issues as these: (i) nature and culture: 'Cultural products and activities become explicit as such only by the use of normative vocabulary that is in principle not reducible to the vocabulary of the natural sciences' [... Brandom 2000a, 33];[38] (ii) the Hegelian pragmatism and his influence about conceptual norms. Kant understood concepts as the norms that determine what we ourselves are responsible for, committed and entitle us to it, by particular acts of judging and acting. However, Kant punted the origins of this normativity into the noumenal realm. "Hegel brought these issues back to earth by understanding normative statuses as social statuses, ... all transcendental constitution is social institution" [... Brandom 2009, 34].
>
> BAVARESCO 2012, 77[39]

Though this social turn in epistemology is progress in terms of its analytic predecessors, it is not connected to Hegel's historicist philosophy. Rockmore's main criticism is that Brandom fails to grasp Hegel's historicism and focuses solely on the constitution of self-consciousness in a society. For Brandom, society is then simply the game of giving and asking for reasons, a social, logical space. For now, what is important is that we recognize that Brandom, although a student of Rorty, remains attached to the semantic goals of the perspective I have developed on analytic philosophy.[40]

1.5 *Conclusions on the Birth of Analytic Philosophy from the Spirit of Hegel*

Since I defend historical relativism, my goal is not to replace the founding-myths of Russell and Moore with "historical truth".[41] Rather, I want to

---

38 Brandom, Robert. *Articulating Reasons: An Introduction to Inferentialism.* Cambridge: Harvard University Press, 2000a.

39 Bavaresco, Agemir. "On Articulating Reasons of Robert Brandom and His Hegelian Methodology." *International Journal of Humanities and Social Science* 2(19), 2012: 74–86. Accessed November 11, 2017.

40 DeVries writes in his comment on Brandom's *A Spirit of Trust:* "How very odd that one of Richard Rorty's best students should end up viewing mind (or spirit) as a mirror of nature" (DeVries 2021b, 239). DeVries, Willem A. 2021b. "Brandom and A Spirit of Trust." *International Journal of Philosophical Studies* 29 (2): 236–250. DOI: 10.1080/09672559.2021.1918822.

41 "Recent work on the origins of analytic philosophy has started to replace the myth with historical truth" (Redding 2007, 7).

historically contextualize the starting point and goals of analytic philosophy. The emergence of analytic philosophy, as I have posited, was based on a myth. Analytic philosophers believed that Hegelianism fundamentally missed the mark by denying that reality exists. Moreover, it is plausible that their rejection of Hegel was influenced by broader cultural forces. Redding writes regarding these broader cultural forces:

> Myths are more than sets of mistaken beliefs about the world, they are cultural products which play constitutive roles in the formation and maintenance of group identities, exemplifying and reflecting back to their members the shared fundamental norms and values binding them as a group.
>
> REDDING 2007, 7

While my project does not intend to offer an exhaustive analysis of specific historical conditions, it engages in reflections on the historical evolution of ideas surrounding Hegel's 'ontological monism'—the belief that only the mind and its concepts exist. In this context, my argument posits that the recent resurgence of interest in Hegel within analytic philosophy represents a critique of the myth of Hegel as an ontological monist.[42] Ontological monism has been an easy enemy, in order to justify logical atomism (see Redding 2007, 8).[43] If we examine the advancements in analytic philosophy through a historical dialectic framework, it appears, however, improbable that the current renewed preoccupation with Hegel as a proponent of semantic holism accurately reflects Hegel's true stance. However, given the analytic line of thinking, a return to a historicist Hegel is not likely. For this reason, Redding's analytic suggestion of how to proceed with our interpretation of Hegel is problematic.

---

42 I agree with Redding on the point that we are dealing with myths of philosophy more than with particular thoughts: "In this sense we treat myths as the central graspable unit of a philosophy that we deal with. We do not attempt to bring philosophies to their historical truth here, but we observe streams of philosophy with regard to their mythological role they play among philosophers. This includes the necessary reduction of philosophies to various slogans as long as we work in a reconstruction of the historical framework, but, of course, this work should not be limited to myths and slogans. The more serious attempts shall be a reconstruction of problems that we face with regard to a systematic approach to the whole" (Redding 2007, 7).

43 It is questionable whether the "monist or speculative, contradiction-embracing logician is the 'real' Hegel" (see Pippin, Robert. *Hegel's Idealism. The Satisfactions of Self-Consciousness*. Cambridge: Cambridge University Press, 1989, 4).

Redding proposes to learn from Russell's mistakes and to adjust our view of Hegel: "Given the fundamental and obvious philosophical errors *known* to lie at the heart of the idealist tradition—that is, those errors learned about from Russell—what *could* be possibly learned from them?" (Redding 2007, 8). It is unclear how to learn from Russell's analysis of idealism because, without Hegel's concept of history, we discuss a strawman of Hegel. In the aftermath of Russell, analytic philosophy directed our attention to the problem of the external world as an anti-Hegelian idea. Rockmore's reconstruction of this line of history introduces the social interpretation of Hegel as a compensation for these former problems. In addition, there might be many more historical and institutional reasons why a return to Hegel occurred. My argument, however, is that the current implementation of Continental thinkers does not respect the problems that are related to Continental theories and thus the divide still exists and it is unlikely that we bridge it with sole efforts of the utilization of Hegel. Most significantly, however, I hold the belief that Hegel's concept of historicity offers a more successful framework for developing a theory of knowledge when compared to Brandom's conceptual realism, which I will examine in the following part.

2 Brandom's Pragmatic Semantics and Its Relation to Realism

Brandom is widely regarded as one of the "founding father[s] of inferentialism" (Veluwenkamp 2021).[44] His combination of inferentialist semantics and normative pragmatics is highly appreciated as "one of the most interesting contemporary initiatives" in demonstrating "what one can do with words" (Marshall 2013, 2). Commentators have praised Brandom's "monumental" *Making It Explicit* in which he "announces" nothing less than "a revolution in the philosophy of language" (Kremer 2010, 227).[45] Hale and Wright describe it as the most "thoroughgoing answer" to the challenge of developing a "broadly use-based" approach to "linguistic competence" (2010, 276).[46] Habermas even

44 Veluwenkamp calls him actually "the founding father" (Veluwenkamp, Herman. "Inferentialist Truth Pluralism." *Ethical Theory and Moral Practice* 24, 2021: 107–120. https://doi.org/10.1007/s10677-020-10145-5).

45 Kremer, Michael. "Representation or Inference. Must We Choose? Should We?" *Reading Brandom. On Making It Explicit.* Edited by Bernhard Weiss, and Jeremy Wanderer. New York: Routledge 2010.

46 Hale, Bob, and Crispin Wright. "Assertibilist Truth and Objective Content—Still Inexplicit?" *Reading Brandom. On Making It Explicit.* Edited by Bernhard Weiss, and Jeremy Wanderer. New York: Routledge 2010.

goes so far as to consider it as a "landmark [Meilenstein]" comparable in importance to Rawls' *Theory of Justice* (see 1999, 138). Brandom's project is, indeed, the ambitious attempt to pragmatically integrate the complete, former dominating paradigm of representationalism through inferential role semantics. Inferential role semantics means that the meaning of an expression depends on its role in inferences that it has through or for other concepts.

Of course, the enormous extent of Brandom's "largely programmatic" endeavor has not only brought praise but also invited strong criticism (Williamson 2010).[47] With regard to its lack of successful application to concrete problems, Williamson, for example, remarks: "if you want an explicit theory of how some particular linguistic construction contributes to the meanings of sentences in which it occurs, the inferentialist is unlikely to have one. Better try the referentialist" (Williamson 2010). He writes dismissively about Brandom's "erudite sophistication and laborious ingenuity" as finally bringing him to the conclusion "that it cannot be done" (Williamson 2010). One can easily see that Brandom's strong rejection of representationalist theories and their 'Cartesian dualism' regarding world and language users has brought him not only friends among philosophers. The harshness of some comments on Brandom's inferentialism makes it seem as if his philosophy has created another divide. In battlefield metaphors, Fodor and Lepore, for example, criticize that Brandom has 'assembled' "an army of supporters" without 'winning' any "battle" (Fodor and Lepore 2007, 691).[48] Their philosophical critique revolves around the argument that inferentialism, which considers the utterance as the smallest unit of meaning, fails to address fundamental questions in semantics, such as how the meaning of a sentence can plausibly be derived from its constituents (see Fodor and Lepore 2001, 465). They view inferentialism as an unsupported "consensus view" without clear evidence and question whether it has truly been demonstrated that inferential role semantics is the "only serious alternative" to an "atomistic referentialism" (2007, 690).

Regarding Brandom's bold claims, McDowell, his colleague and friend in Pittsburgh, raises the question of whether his "strong inferentialism" needs to break so radically with all representationalist positions, especially considering

47 Williamson, Timothy. "Review of Robert Brandom, Reason in Philosophy: Animating Ideas." *Times Literary Supplement*, 5579, 2010: 22–23. http://media.philosophy.ox.ac.uk/assets/pdf_file/0004/35833/tlsbrandom.pdf.

48 Fodor, Jerry and Ernie Lepore. "Brandom's Burdens: Compositionality and Inferentialism." *Philosophy and Phenomenological Research* 63 (2), 2001: 465–481. doi:10.1111/j.1933-1592.2001.tb00119.x.

that his approach can include non-linguistic elements (1997, 159).[49] Brandom himself suggests, albeit buried in a footnote, the possibility of a third, combined approach alongside inferentialism and representationalism (see McDowell 1997, 158–159; Brandom 1998, 669).[50] In light of the stark division between inferentialists and representationalists, McDowell expresses therefore his "suspicion" that Brandom might ultimately engage in a battle against a "chorus of strawmen" (1997, 159). Certainly, it can be quite a surprise for a theorist who employs representationalist tools and has found some practical success in explaining the meaning of sentences when he reads that Brandom has delivered a final and, as Taylor describes it, "devastating critique of the atomism which is implicit in the mainstream post-Cartesian epistemology" (2010, 32).[51] However, in response to the strong reactions from both sides, Murzi and Steinberger propose an alternative approach to the referentialism/inferentialism divide.[52] They specifically address Williamson's accusation that Brandom's inferentialist semantic theory is an "untested, and mostly under-developed, alternative inferentialist semantic theory" (Williamson 2010).[53] According to them, Williamson's sharp criticism of inferentialism stems from a conflation of meta-semantics and semantics. An inferentialist could, according to their distinction, turn "to a referential semantic theory" when dealing with "the ordinary semantic concepts of reference" that refer to "extra-linguistic items as semantic values" (Murzi and Steinberger 2017, 215). Simultaneously, the inferentialist can still fully embrace the meta-semantic "spirit" of inferential role semantics (Murzi and Steinberger 2017, 215). In this context, meta-semantics concerns the broader and metaphysical question of how an expression acquires meaning overall, while the conventional semantics of ordinary concepts can still be

49 Strong inferentialism can allow for 'narrow' and 'wide' inferences. Veluwenkamp gives some examples: 1. Narrow inferences: 'x is red', therefore 'x is colored', therefore 'x is scarlet', therefore 'x is red'. 2. Wide inferences: red things, therefore 'x is red' (see Veluwenkamp 2021).

50 Brandom writes: "Other possibilities include treating neither representation nor inference as explanatorily prior to the other. One might then go on to explain both in terms of some third notion, which is treated as more fundamental" (Brandom 1998, 669).

51 Taylor, Charles. "Language not Mysterious?" *Reading Brandom—On Making It Explicit. Edited by* Bernhard Weiss, and Jeremy Wanderer, 32–47. New York: Routledge 2010.

52 Murzi, Julien, and Florian Steinberger. "Inferentialism." *A Companion to the Philosophy of Language.* Edited by Bob Hale, Crispin Wright, and Alexander Miller. Malden, Oxford: Wiley Blackwell, 2017. https://doi.org/10.1002/9781118972090.ch9.

53 Williamson also criticizes that Brandom's programmatic approach is attracting only readers who "prefer philosophy to come in vast, vague programmes, like the election manifestos of parties that know they will never have to govern" (Williamson 2010).

employed alongside it (see Murzi and Steinberger 2017, 215).[54] For example, a representationalist could plausibly explain how a sentence like 'the cat is on the mat' is learned by initially becoming familiar with its constituents.[55] Therefore, when Brandom characterizes his approach as "self-consciously one-sided," it does not imply a rejection of these concrete representationalist solutions to such problems. Rather, it indicates that he engages with the question of meaning at a meta-semantic level, specifically from a holistic perspective (Brandom 2007, 653).[56]

Besides the replacement of representationalism, Brandom's major motivation is to respond to the strong meta-semantic objection that pure inferentialism will "inevitably 'lose the world'" (Brandom 1998, 331). Because inferentialism does not directly anchor its system in a mind-external object, but instead explains the meaning of a concept through the meaning of another concept, there is a risk of it descending into global relativism. Given that the accusation of being a global relativist remains unacceptable to many philosophers, Brandom's objective goes beyond merely replacing representationalism in a meta-semantic sense. He aims to demonstrate and prove that inferentialism, in the form of normative pragmatics, can offer an objective notion of how we can improve our understanding of the mind-external world.

Now, the question arises as to whether the notion of a mind-external world, in contrast to our linguistic objects, reintroduces elements of representationalism. Rydenfelt, for instance, points out that with this objective understanding of the mind-external world, Brandom positions himself "one step to the right from full-fledged anti-representationalism" (Rydenfelt 2019, 2).[57] Expanding on this observation, Rydenfelt explains that in Brandom's inferentialism "'ordinary empirical claims' are *about* things, and not merely expressive of our attitude" (2019, 13). This inherent aboutness of assertions potentially conflicts with

---

54 The discussion has some twists and turns since, as Veluwenkamp points out, the restriction of a theory to a certain domain still faces problems that these "local theories" can look very similar, "while their meaning is determined in fundamentally different ways" (Veluwenkamp 2021).

55 Indeed, inferentialism employed for explaining the process of learning would sound absurd: we would first learn sentences as a whole package and afterward explain and understand their constituents. However, as much as earth could still be described as the center of the universe in an insanely complex model, it is probably also possible for explaining theories of learning through such an inferentialist approach. It should be clear, however, that this would be highly impractical.

56 Brandom, Robert. "Inferentialism and Some of Its Challenges." Philosophy and Phenomenological Research LXXIV (3), 2007: 651–676. DOI 10.1111/j.1933-1592.2007.00044.x.

57 Rydenfelt, Henrik. "Realism without Representationalism." *Synthese* 198, 2019: 2901–2918. https://doi.org/10.1007/s11229-019-02251-4.

pure inferentialism, where Brandom asserts that "truth and reference are philosophers' fiction" (1998, 324).[58]

Therefore, my intention is not to assess Brandom's success in confronting the "explicit enemy" of representationalism (Fodor and Lepore 2001, 467). Undoubtedly, the polarization evident in the inferentialism/representationalism[59] debate indicates that Brandom has had a significant impact on the community of analytic philosophers. However, my objective is to demonstrate that Brandom's inferentialism still relies, possibly unintentionally, on a mind-external, metaphysical realism that is fundamentally incompatible with Hegel's idealism.

However, the matter is more nuanced than it appears. Brandom introduces his mind-external realism within the framework of inferentialism, contending that it constitutes a crucial element in our communicative practices and can be rationally justified. It is essential to recognize that, if Brandom successfully executes his proposed strategy as articulated in *Making It Explicit*, it may not constitute a full-fledged metaphysical, mind-external realism. Instead, it would signify a more modest pragmatist claim regarding mind-external reality as a procedural condition that language users must transcendently presuppose to engage in mutual interaction. While this procedural realism still diverges from Hegel's idealism by delineating content and form, it could potentially counter the relativism I advocate.

In order to give a more precise answer to the question of the compatibility of Hegel and Brandom's conceptual realism, I will first reconstruct Brandom's

---

58 With regard to the boldness of Brandom's overall approach, Huetter-Almerigi provides an interesting interpretation. After his publication of *Making It Explicit*, Rorty recommended being "more arrogant" against possible "accusations of relativism" (Rorty 1997b, 177; Huetter-Almerigi 2020, 10). As Huetter-Almerigi concludes "inferentialism, even in Brandom's causally constrained form, will never be enough to convince people with strong realist intuitions [... so] he might as well go with revolution" (Huetter-Almerigi 2020, 10). With his "immodest metaphilosophical framework" (Rorty 1997b, 177), Brandom has therefore decided, against Rorty's advice, and defends the route of procedural realism. According to this procedural realism, we get closer to reality (Huetter-Almerigi, Yvonne. "Two Forms of Realism—Making Sense of Rorty's Controversy with Brandom and Ramberg over Objectivity." *European Journal of Pragmatism and American Philosophy* [Online], XII-1 2020:1–14, URL: http://journals.openedition.org/ejpap/1868).

59 I use the terms 'referentialism' and 'representationalism' interchangeably in this context. It's important to note that, according to referentialism, words may not necessarily involve mental images, unlike representationalism, which often emphasizes the role of mental representations. The central point in the debate, however, revolves around the reference to an external world, a common aspect shared by both representationalism and referentialism.

crossover of inferentialist semantics with normative pragmatics. As he developed his approach mostly in his first major publication *Making It Explicit*, I will first focus on this book (part 2.2). Afterward, I will look at how he attempts to transform Hegel and adopt his ideas in his book *A Spirit of Trust* (part 2.3 and 2.4). Moreover, there are some more general points regarding his inferentialist strategy that I need to explain before I can begin my reconstruction.

In his influential work, *Making It Explicit*, Brandom endeavors to address the Sellarsian problem of inferentialism as an empty coherentism, as discussed previously (see point 2.1.3). He aims to demonstrate that the combination of the inferentialist approach with normative pragmatics can meet the requirements of realism. To accomplish this, he employs a formal inferentialist framework which he terms 'linguistic rationalism'. For Brandom, 'linguistic rationalism' means first and foremost that "semantics must answer to pragmatics" (Brandom 1998, 83). This approach builds upon the traditional integration of propositional reference-semantics into use-based approaches, but in Brandom's case, it takes the form of an inferentialist, normative pragmatics. However, Brandom goes beyond mere pragmatic integration and seeks to show how this wholly pragmatic, inferential structure can be objectively grounded from within discourse. According to his immanent strategy, speakers maintain an implicit connection to a mind-external reality. Rather than assuming mind-external reality as a mystical given, it is produced reliably as part of their perspective-transcending communication. They achieve this by demanding from one another (and themselves) to make explicit what their utterances fundamentally concern, namely, mind-external reality. Brandom firmly believes that speakers engage with such a mind-external reality through their practice of score-keeping, which transcends their individual perspectives. In this sense, we can understand that *Making It Explicit* not only aims to make certain conditions of our utterances explicit but, at its deepest level, strives to make the objective norm explicit through which we ultimately approximate mind-external reality. It is crucial to recognize that making mind-external reality explicit does not entail arriving at a final and definitive ontological concept of it. Rather, it involves defining the procedural norm in our communication that ensures conceptual progress toward it.

To summarize, Brandom, following the principles of inferential semantics, ambitiously seeks to integrate reference-semantics pragmatically. Moreover, he aims to safeguard his approach from the pitfalls of relativism and provide a demonstration of objectivity. To accomplish this, he employs a wide range of sophisticated inferentialist tools that align with the analytic tradition. However, things become even more intricate as Brandom, following *Making It Explicit*, introduces a Neo-Hegelian model into the analytic discourse, incorporating the complex Hegelian/Continental terminology. Brandom argues that the conceptual realism he presents in *Making It Explicit* was already developed but overlooked in Hegel's 'objective idealism'. From Brandom's perspective, Hegel's philosophy offers a highly refined form of inferentialism that can complement his own ideas, particularly concerning its practical manifestation within developing societies.

Now, while Brandom draws on Hegel's ideas, it is important to note that his approach does not entail a complete embrace of Hegel's idealism. This is because Brandom's utilization of Hegelian ideas occurs within the framework of what I perceive as his analytic philosophy. As such, Brandom's philosophy primarily engages with figures from the analytic tradition, while incorporating arguments from Hegel into the ongoing debate. As McDowell suggests, Brandom "motivates his approach, to a large extent, by its competitor's flaws", i.e., the representationalism of the analytic tradition (McDowell 1997, 157). His project operates therefore on the grounds of a critique regarding representationalism, particularly, the representationalism that results from the early, analytic philosophy. In my view, it eventually inhibits a genuine integration of Hegel's ideas, which, however, is also not Brandom's intended objective.

I need to specify the impact of the analytic tradition on Brandom's integration of Hegel a bit more. The major problem of the early analytic tradition is for Brandom that it failed to provide a successful account of atomic knowledge in order to represent extra-linguistic, mind-external reality. In response to these problems of developing a representationalist theory of meaning, Brandom, therefore, reverses the former Augustinian strategy to "work up from [...] words to [...] sentences, and only then [...] to consider the structure of the language-game" (McDowell 1997, 158). However, even though this controversial reversal (see McDowell 1997, 158) represents a break with the early analytic "bottom-up" tradition, it does not include a radical revision of early analytic philosophy's emphasis on mind-external realism (Brandom 2007, 650). It is, however, this mind-external realism that is so strongly opposed to Hegelian idealism. My major thesis is therefore that the maintained frame related to mind-external realism conflicts with the integration of Hegel's ideas into conceptual realism. It conflicts, even though Hegel as a sort of 'holistic' thinker can be roughly grouped into the Brandomian contextualist camp of philosophers who operate "top-down" (Brandom 2007, 651). Nevertheless, as indicated in the introduction, it is not only that it inhibits the integration of Hegel. In my mind, it also falls behind the achievements of the Hegelian concept of historicity by reverting to metaphysical realism.

To specify the realist frame, early analytic philosophy wanted to demonstrate, first, explicitly against 'Hegelianism' that reality exists—a claim that

Hegel never denied. Second, analytic philosophers attempted to prove that we are in immediate contact with this mind-external reality. For Hegel, such proof is unlikely and given his, as I argue, relativist approach is not necessary. Now, in the spirit of Sellars' 'linguistic rationalism', Brandom responds to this analytic tradition in two ways. First, he wants to remove unintelligible immediacies, which as a first step is directed against analytic representationalism and its so-called 'myth of the given'. As a consequence, however, it leads to a neglect of the experiential themes in Hegel's *Phenomenology*. This is because, for Brandom's linguistic rationalism, the term 'experience' carries the bad taste of a 'given', or better said, of non-inferential knowledge. Second, he wants to respond to the accusation of empty coherentism, the problem of 'bad' idealism. This means that something mind-external must constrain our communication. This requirement, which Brandom takes very seriously, can be associated with early analytic philosophy's anti-idealism.[60] As a consequence of this maintained frame, Brandom must also interpret Hegel's idealism as related to a mind-external reality. Through this lens, Hegel's 'objective idealism' is not a relativist, "frictionless spinning in the void" (see McDowell 1996, 42). This interpretation, however, stands in strong contrast to my view of Hegel's idealism as a relativist historicism, which I evaluate to be the better philosophy.

The question is whether it is genuinely possible to translate Hegelian ideas into the analytic tradition while preserving a realist framework. Brandom's conceptual realism, with its holistic structure, appears to resemble an idealist framework, suggesting a potential association of his ideas with Hegelian philosophy. However, in his appropriation of Hegelian concepts, Brandom overlooks the inherent historical relativist structure in Hegel's idealism.

In the endeavor to reclaim Hegel from the originally anti-idealist tradition of early analytic philosophy, Brandom reinterprets Hegel as a semantic realist. Nevertheless, this adaptation faces challenges due to Hegel's rejection of transcendental, mind-external realism, as well as procedural realism that separates form and content. Moreover, while Brandom's theory emphasizes the relationship of concepts to the mind-external world, Hegel primarily grounds his understanding of knowledge on the mind-internal standard of experience and its associated reflection, as I will elaborate on in Chapter Four. This inherent difference between the two perspectives underscores their incompatibility. Therefore, the legitimate question arises: Can Brandom's inferential role semantics genuinely harmonize with Hegelian concepts? As I see it, Brandom

60 Analytic anti-realists are more likely to soften or even drop this position. This tendency includes, for example, Putnam, Davidson, and Rorty (see Huetter-Almerigi 2020).

and Hegel emerge as strict competitors, and the endeavor to merge Hegelian ideas into the analytic tradition while maintaining a realist frame is likely impossible due to the disparities in conceptual realism and the relativist concept of experience.[61]

I will explain the details of the historicist relativism of Hegel's *Phenomenology* in Chapter Four. Here, I hope that the brief introduction of my interpretation of Hegel clarifies my specific perspective on Brandom's philosophy. One could say, the major reason for my rejection of Brandom's *Making It Explicit* as Hegelian lies in its negligence of the rather Continental experiential motives of a Hegelian mind-internal, historicist phenomenology. This seems to change in part with Brandom's publication of *A Spirit of Trust,* a more than 800-page interpretation of Hegel's *Phenomenology of Spirit* that was published almost 30 years after *Making It Explicit.* While Brandom does not use the concepts of experience and history in his early works, in his later book *A Spirit of Trust*, he introduces an explicit concept of experience that, as he claims, is borrowed from Hegel. However, he now understands experience as fundamentally grounded in 'attitude-transcending' things-in-themselves, which he interprets as 'noumena'. In other words, Brandom introduces us here to an uncommon reading of Hegel's term 'in-itself' [*an-sich*]. The 'in-itself'—or, according to Brandom's surprising interpretation, the noumenon—is the stable entity that we approach through a historical process of knitting and tightening a semantic net. The relationship between experience and the 'in-itself' is then supposed to ground Hegel's historicist idealism as a conceptual realism. From Brandom's perspective, Hegel, despite his rejection of philosophy as edifying, occurs now as a "prophet" who delivers the model for approaching a better future. The stark contrast to my position therefore remains: Hegel focuses on relative, historical, human experience without an explicit relationship to noumena or a necessary, teleological 'happy end' of human history. I therefore also regard Brandom's later account—although he dives deep into Hegel's *Phenomenology*—as opposed to Hegel's philosophy. The general problem for Brandom's account, as I will reconstruct, is that if experience has no relationship to things in themselves, it would deliver only a relativist ground for knowledge. For the conceptual realism that Brandom pursues, this is not enough.

Given the order of my interpretation, I will first discuss Brandom's earlier approach presented in his groundbreaking book *Making It Explicit.* This account does not yet include a concrete concept of experience and is not yet

61 Rockmore has developed these arguments for a phenomenology of experience in several books (see, for example, Rockmore, Tom. *Kant and Phenomenology*, London and Chicago: University of Chicago Press, 2010).

explicitly Hegelian. I will focus on the intended procedural realism that he, as I see it, conflates with mind-external realism. In the subsequent point, I will then discuss his later approach to Hegel and his integration of the concept of experience based on the concepts of noumena and phenomena.

In accordance with Brandom's account in *Making It Explicit,* there are three characteristics that stand in strong contrast to Hegelian philosophy:

1 his mind-external realism related to his semantic inferentialism
2 his procedural realism that centralizes the concept of a propositional assertion
3 his formalized, linguistic model that includes different proofs of objectivity.

In the following points, I will reconstruct Brandom's *Making It Explicit* highlighting these three characteristics.

2.1 *Mind-External Realism and Inferentialism*

Mind-external, also called metaphysical realism, is the idea that there are mind-independent, determinate things that can be known. On the contrary, procedural realism is the idea that we do not need to know mind-external things in order to secure objective methods for our communities. It means, for example, that the acknowledgment of practical reasons can lead to an objective set of regulative or moral norms. Norms can accordingly be achieved without the involvement of mind-external facts. Brandom's conceptual realism is intended to secure objective practices that can establish an approximation toward mind-external reality without relying on a given mind-external reality. In my view, however, his inferentialism remains a mind-external, metaphysical realism that only utilizes procedural realist methods. To put it simply, our normative discourses, according to Brandom, depend in a determinate way on how things really are.[62] In the following points, I will attempt to carve out this mind-external realism from his inferentialist theory. I first start with a general description of his account and will work on the subsequent points toward his crucial proofs of objectivity.

---

62 From my perspective, the notion of a mind-independent reality appears to be an abstract concept that we create to articulate our experiences. It is crucial to recognize that this idea of a mind-independent reality is inherently a product of human construction, making it a form of mind-dependent mind-independence. I intend to delve deeper into this matter, especially in relation to Hegel, later on. It is worth noting that Brandom seems to acknowledge this predicament. However, he appears confident in having discovered a solution to secure a path toward mind-external reality through procedural realism.

Brandom's approach is commonly described as inferentialism, which is supposed to liberate words like 'true' and 'refers' from the representationalist paradigm (see Hookway 2016).[63] His major intent is to translate the more common representationalist vocabulary into procedural, inferentialist terms of normative correctness (see McDowell 1997, 157). 'Inferentialism' means that all concepts are understood with regard to other concepts, in Brandom's case, particularly with regard to the role they play in the human, social practice of 'giving and asking for reasons'. 'Pointing at something', as an example for representing something mind-external, would accordingly be understood through the inferential practice of a human exchange of reasons. The gesture itself does not have a meaning that we can merely understand intuitively. According to inferentialism, 'pointing' is rather explained as a reason for someone looking in a certain direction. It, therefore, is part of a larger structure of meaning within a particular language game. Brandom is motivated by the shortcomings of representationalism to explain these more complex structures of meaning. He summarizes: "the representational model, afterall, does not purport to tell us about justification [...]; its claim is that the use of our empirical vocabularies stands in normative semantic relations to the world, in that how things are determines the correctness of our claims in the sense of their truth" (Brandom 2000b, 161).[64] As he suggests, there is no simple semantic relation between words and the world. However, while he wants to explain the representationalist vocabulary by only inferential terms, a significant remainder of mind-external realism is preserved in Brandom's theory. Brandom's major goal, as I see it, is to prove an inferentially mediated contact with a stable, extra-linguistic, mind-external reality. According to him, the inferential structure depends on how things are. He expresses this relationship between the objectivity of our discourse activity and a mind-external reality often. The following statement emphasizes 'real things' and their role in our discourses:

> The objectivity of conceptual norms requires that any attitude of taking, treating or assessing as correct an application of a concept in forming a belief or making a claim be coherently conceivable as mistaken, because of how things are with the objects the belief or claim is about.
>
> BRANDOM 1998, 63

---

63 Hookway, C.J. "Pragmatism." *The Stanford Encyclopedia of Philosophy*. Edited by Edward N. Zalta. 2016 Accessed on November 25, 2017. <https://plato.stanford.edu/archives/sum2016/entries/pragmatism/>.

64 Brandom, Robert. "Vocabularies of Pragmatism: Synthesizing Naturalism and Historicism." *Rorty and His Critics*. Oxford: Blackwell Publishers, 2000b.

In other words, the objectivity of our discourses—our practice of giving and asking for reasons—depends on real, mind-external things. The presupposition of a mind-external reality and its power over our discourse looks simple, but Brandom embeds it in his more complex inferentialist model. Given this approach, we must take into account that all semantic content is normatively charged through social interaction. Brandom's semantic concept of a mind-external reality, therefore, does not mean that he simply introduces an ontologically different sphere of mind-external objects as truth-makers. The situation is more complex. However, to understand his supposedly non-dualistic, inferentialist notion of mind-external reality, we need to further reconstruct the basic ideas of his approach.

Brandom attempts to explain the semantic content of our statements by a pragmatic integration and thereby replaces the 'older' paradigms of consciousness and the mind-external world with pragmatics.[65] He summarizes the overall intention with regard to this pragmatic integration: "the book is an attempt to explain the meanings of linguistic expressions in terms of their use" (1997a, 153).[66] In order to fulfill the demand of 'semantics answering pragmatics', the book starts with "an account of social practices" (Brandom 1997a, 153), investigating the structural conditions that make them "linguistic practices" (Brandom 1997a, 153). For this account, Brandom distinguishes the capacity to speak from the capacity to only produce sounds. Speaking is about our mastery of treating something inferentially, instead of merely referring to things. For example, parrots do not understand the concept of red (see Brandom 2009, 170).[67] If they 'answer' 'I see red', it will only be a stimulus-response. Humans, on the contrary, can master this concept by understanding its inferential role (See Mosteller 2014, 81).[68] Understanding its inferential role means accepting that an utterance has certain implications for a system of utterances. For example, 'red' means that it stands in inferential relations to similar concepts like 'green' and 'orange', but also to more universal concepts like 'color'. Such a "conceptual role semantics" focuses on the question of how we produce content by embedding concepts into a net of inferences (see Brandom 1997a, 153).

65 Marshall shares the same opinion: "It showed how Brandom's work emerged out of the broad shift in the philosophy of language from semantics to pragmatics that also informed speech theory" (Marshall 2013, 1).

66 Brandom, Robert. "Review: Précis of Making It Explicit." *Philosophy and Phenomenological Research* 57 (1), 1997a: 153–156. http://www.jstor.org/stable/2953784.

67 Brandom, Robert. *Reason in Philosophy: Animating Ideas.* Harvard University Press, 2009.

68 Mosteller, Timothy M. *Theories of Truth: An Introduction.* London: Bloomsbury Publishing, 2014.

Content is not magically picked up and introduced into a formal system of language, but it is produced by the actions of the speakers and how they position the related concepts in an inferential net.

Through the notion of an inferential net, Brandom explicitly seeks to eliminate any reliance on a 'mystical' given, as we previously discussed in relation to Sellars. Consequently, his account aligns with the traditions of analytic philosophers who eschew metaphysical matters unless they can be justified and explained within linguistic frameworks. Nevertheless, although Brandom states "I do think of myself as an analytic philosopher" (Williams 2013, 370), he does not unequivocally dismiss all Continental, non-linguistic considerations as meaningless. In his response to Taylor, he emphasizes his adoption of a broad and encompassing perspective:

> I am working within a broadly pragmatist tradition that includes not only the classical American pragmatists but also the early Heidegger and the later Wittgenstein. My pragmatism is distinguished from theirs, however, by being joined to a kind of philosophical rationalism. It is this unusual combination of commitments that draws me to Hegel.
>
> BRANDOM 2010, 301[69]

In particular, he believes that the Continental philosophy has a stronger tendency towards holism that, in a sense, uses concepts of inferentialism. Despite this open and valuable commitment to such a broad spectrum of philosophies, Brandom's conflict with Taylor consists in his insistence on the assertion as the central part of our linguistic activity.[70] This focus on assertions distinguishes Brandom from Continental thinkers like Heidegger, Wittgenstein, and Hegel, and probably even from the pragmatists. In this spirit, he also describes the assertion as the downtown of language (2010, 301).

Assertions are central to Brandom's conceptual realism. It represents a controversial point since the pragmatic turn in analytic philosophy moved away

69 Brandom, Robert. "Reply to Charles Taylor's 'Language not mysterious?'" *Reading Brandom. On Making It Explicit.* Edited by Bernhard Weiss, and Jeremy Wanderer, New York: Routledge 2010.

70 Nevertheless, against the atomism of "the mainstream post-Cartesian epistemology", Taylor emphasizes the importance of the pragmatic turn that also focuses on a so called multi-dimensional holism that he finds in Brandom (Taylor 2010, 32). Language is for Taylor more complex than simply inventing words and labeling ready-made objects. It is not only an instrument, or a labeling-machine. Taylor dismisses also the idea that we take in "pieces of information" (see Taylor 2010, 32) and bulk them together in a bucket of experience. He points out that Brandom makes important advances on these aspects.

from this focus. Austin's initial distinction between performatives and constatives (roughly an assertion), for example, turned out to be too rigid. Brandom acknowledges the fact that assertions are embedded in performative actions, which constitutes the normative, pragmatic part of his approach. Yet, his overall goal is to show the objective meaning of our linguistic activity, contrary to Austin, through the proposition expressed by assertions. Using an assertion with its propositional content is ultimately the ground for his objectivity proof and builds a pillar of his inferentialism. Nevertheless, the use-theoretic approach also remains a pillar of his overall normative approach. So, for example, normative vocabulary can be applied in cases when we cannot specify the content of an utterance with semantics alone. In these cases, we do not have to understand what the statement is about, but what we do when we use an assertion. Brandom mentions in this regard simple cases of use specification. So, for example, we can use "the word 'not' to express negation" (Brandom 1997a, 153). In this case, there is little to no doubt that we can specify the use of a word in semantic terms. We find more difficult cases in the field of theories that present concepts as very vague. These are, for example, "descriptions of the movements of particles expressed in the vocabulary of physics" (Brandom 1997a, 154). The use of such vocabulary cannot support a stable meaning, since "expressed by various noises or inscriptions, it will, in general, fail to settle even that anything is meant or expressed by them" (Brandom 1997a, 154). According to Brandom, this difficulty in settling specific pragmatic meanings in assertions is because our normal "intentional vocabulary", as well as our "naturalistic vocabulary", cannot transfer what we mean entirely (Brandom 1997a, 154). In order to solve these difficulties of settling the use of a word in assertions, Brandom wants to apply "normative vocabulary" (Brandom 1997a, 154). In other words, the content of an utterance depends on what we oblige ourselves to do when we make these utterances.

As already discussed, for Brandom, each communicational practice includes norms of correctness with regard to the used expressions. In other words, speech acts like assertions need to fit social standards (see Brandom 1997a, 154). The limits of intentional vocabulary and natural vocabulary are found where we do not speak about clearly observable phenomena or describable objects of physics. In these cases, we must become aware of how we are doing things with language. Here, Brandom suggests that our assertions are normative and that the meaning of such extreme physical examples can be more correctly addressed with normative vocabulary. Nevertheless, this does not change for him the fact that the normative vocabulary depends on the propositional structure of assertions because only they can produce a structure according to which we can exchange reasons. Ultimately these assertions depend on how things really are, as he writes:

> The properties governing the application of those concepts depend on what inferences involving them are correct, that is, on what really follows from what. And that depends on how things really are with electrons and aromatic compounds, not just on what judgments and inferences we endorse.
>
> BRANDOM 2000a, 27

According to Brandom, our normative vocabulary somehow expresses what is real about our assertions related to "electrons and aromatic compounds" (Brandom 2000a, 27). The content of assertions is not semantically 'substantialized', but expressed through the normative 'truth-making' practice of tightening a semantic net within our continuous discourse (Brandom 2000a, 27). The problem, however, is that we cannot simply introduce an ontological concept of reality. This is because inferentialism does not allow for unexplained outside events. We therefore must understand how we address mind-external reality with our statements by analyzing our normative involvement with the world and society. To unravel and explain this notion Brandom cannot construct a short argument. Overall, *Making It Explicit* is supposed to create a 741-page pathway towards such proof of the objectivity of our discourse, i.e., to explain the determinate relationship between our discourses and the mind-external world. I will therefore look at the structure of the book now.

In Chapter One, Brandom introduces his concept of "normative pragmatics" and explores its roots in the works of Kant, Frege, and Wittgenstein (see Brandom 1997a, 154). His aim is to provide an account of meaning without relying on naturalist vocabulary. In this context, Brandom assumes that all speech acts are normative, stating that: "interpreting states, performances, and expressions as semantically or intentionally contentful is understood as attributing to their occurrence an ineliminably normative pragmatic significance" (1997a, 154). In addition to the fact that every statement is normative, Brandom treats this implicitly normative practice as in principle explicable. The reason for this is that these norms are "instituted by social, practical activity" (1997a, 154). At its core, this social activity involves tracking commitments and entitlements (see Brandom 1997a, 154). He refers to this discursive practice as 'score-keeping':

> Beginning with basic deontic score-keeping attitudes and the practices that govern them, an account is offered of how locutions must be used

> in order to express explicitly the very normative notions—is committed, is permitted, ought, and so on—that are appealed to in laying out the normative pragmatics.
>
> BRANDOM 1997a, 154

As indicated by Brandom, the underlying basis of our practice of score-keeping is rooted in the locution and its propositional content (Brandom 1997a, 154). The terms "attitudes of attributing" and "acknowledging deontic statuses of commitment and entitlement" are key to understanding how we integrate these locutions in social contexts (Brandom 1997a, 154). They represent the fundamental actions by which we exchange the propositional content of locutions and hold each other accountable. Brandom's idea is that we attribute normative status to each other which means that we hold each other accountable according to the locutions that we commit to. To put it in normative terms, we take over responsibility and expect others to be responsible with regard to what they say. The dialogical practice of score-keeping is then primarily driven by the tracking of commitments and the continuous assessment of the logical consequences that arise from commitments. By evaluating the propositional content of locutions, we can determine their compatibility or incompatibility, allowing us to maintain a record of the justified contents we believe in. Brandom's pragmatic approach provides an account of knowledge that relies on a shared score-keeping system. In this context, it is significant that Brandom's pragmatics maintains a connection to the semantic notion of propositions, as he states:

> This is a matter of moving from pragmatics to semantics. The defining characteristic of discursive practice is the production and consumption of specifically propositional contents.
>
> BRANDOM 1997a, 154

Although Brandom advocates for a perspective where knowledge is "norms all the way down" and emphasizes the move from semantics to pragmatics, in this instance, he takes a step back and shifts from pragmatics to semantics (Brandom 1998, 44). Propositional content, alongside the normative charge of every claim, is a crucial element of his communication system. However, this raises the question of whether this step back contradicts the original aim of achieving a complete pragmatic integration of reference. If propositions are not dependent on the speakers' attitudes but rather on the 'things' they 'really' stand for, it appears to conflict with the claim that things are normative all the way down. Nonetheless, Brandom justifies the necessity of this step back to semantics in Chapter Two, where he argues that "propositional contentfulness should be understood in terms of inferential articulation; propositions are what can serve as premises and conclusions of inferences, that is, can serve as and stand in need of reasons" (Brandom 1997a, 154). In other words, without propositions, there would be no inferences, and without inferences, the social practice of giving and asking for reasons would not exist. Moreover, without the exchange of reasons, we could not consider ourselves rational. Therefore, propositions are a fundamental component of Brandom's framework. Specifically, propositions serve the purpose of ensuring consistent compatibility with other propositional statements or indicating incompatibility. They play a crucial role in establishing a shared foundation that enables reliable, rule-based exchanges of reasons between different speakers within a pragmatic discourse. Propositions, in this sense, provide a common surface that facilitates rational beings to make commitments to their assertions and verify entitlements to commitments within their communication community.

In Chapters Three and Four, Brandom delves back into the pragmatic aspects by further examining the activity of giving and asking for reasons, focusing on the involved speakers and their normative status within the discourse. He addresses various problems in the philosophy of mind, which are not relevant to our discussion here. It is interesting to note that Chapter Five, which explores the traditional function of truth, is not mentioned in Brandom's own summary of the book (Brandom 1997a, 155; Brandom 1998, xv). Essentially, Brandom's position is that the concept of truth cannot fully account for the meaning of a sentence without being interpreted through the lens of inferentialism. Nevertheless, it is worth mentioning that Brandom's approach to truth is not entirely deflationary, meaning he does not discard the notion of truth altogether like Rorty, who seeks to "get rid of both [...] 'truths' and 'facts'" (Rorty 2000, 183). Brandom aims to go beyond Rorty's focus on the justification of our beliefs through other beliefs. Instead, he seeks to establish a modified relationship with facts that serve as truth-makers in a certain sense. To clarify, Brandom considers facts as true claims. However, as Reinikainen emphasizes, the central concern lies in our actions and not in providing a semantic account of the truth predicate itself (Reinikainen 2021, 60). I will provide a more detailed explanation of this understanding of truth later (see point 2.2.3). For now, it is sufficient to say that truth is a term with a strong representationalist connotation, and preserving the concept of truth can potentially conflict with inferentialism.

So far, we have outlined the three crucial steps that form the basis of Brandom's inferential, conceptual realism. Firstly, we established that meaning is normatively dependent on use. Secondly, we recognized that use is reliant on inferences. And thirdly, we acknowledged that inferences depend on propositions, as they involve treating something as contentful across different utterances and perspectives. This third step sets the stage for Brandom's main objective: to demonstrate how the practice of giving and asking for reasons, with its propositional content, connects to a reality beyond the mind. To support this, Brandom discusses the inferential integration of deictic vocabulary in Chapters Six and Seven. Specifically, Brandom aims to show that "substitutional inferences" can account for singular terms and predicates, as these terms cannot "directly play the inferential role of premise or conclusion in an argument" (Brandom 1997a, 155). Since utterances are the smallest unit of meaning, inferentialism faces challenges when dealing with components below the 'sentential level' (the level of sentences). So, the question must be answered of how we can understand the smaller components of an utterance. In this regard, Chapter Six explores the notion of anaphora which is "the relation between a pronoun and its antecedent" (Brandom 1997a, 155). The concept of anaphora is proposed as a means to explain singular terms by situating them within our linguistic framework through inferentialist terms. In Chapter Seven, Brandom argues that anaphora can even capture unrepeatable, deictic components like "demonstrative tokens" (Brandom 1997a, 155). Thereby, they anchor our pragmatic, otherwise subjective discourses in a shared reality. The substitutional, anaphoric account developed in these chapters establishes the foundation for securing objective communication, illustrating how speakers can engage with the content of other speakers by substituting elements of their speech with their own terms. I will provide a more detailed explanation of this later on.

Chapter Eight will finally address Brandom's main target, which is to demonstrate how his inferentialist model "institute[s] objective norms" of discourse that, moreover, have to be directed towards attitude-transcendent facts, i.e., a mind-external reality. This involves his well-known distinction of 'de dicto' and 'de re' descriptions and ascriptions. Both of these terms describe our capability to evaluate content regarding the perspective of the speaker uttering it and transforming it into a perspective that attempts to transcend this perspectival attitude-dependence.

Brandom's proof regarding the correctness of procedural conditions aims to ultimately establish a formal and procedural understanding of the mind-external world. In the context of dialog practice, this procedural understanding of the mind-external reality must be the same for everyone. According to Brandom, it must ensure the generation of meaningful content that the entire "linguistic community may be wrong about" (Brandom 1997a, 156). By acknowledging the potential for error, he aims to transcend the limitations of the 'we-perspective', which asserts that a community is incapable of generating truths beyond its own perspective. Nonetheless, this poses a challenge in terms of rectifying false beliefs. Brandom's central and pivotal argument suggests that discourses are guided by substitutional inferences, wherein speakers assess the coherence of the statements of others. Through this method, which allows for the objective rejection of incompatible propositions, an unbiased corrective approach emerges. As a result, engaging in the correct practices enables us to establish a normative connection to external reality, focusing on methodology rather than essence.

Brandom's inferentialism offers a unified framework for addressing representationalist challenges in analytic philosophy. A closer examination of inferentialism is necessary to understand its implications. Within inferentialism, a distinction can be made between logical and material inferentialism. Logical inferentialism aims to explain logical constants and operators through inferences. For instance, the logical operator 'and' can be fully explained using introduction and elimination rules, as per "Gentzen-style semantics" (Fodor and Lepore 2007, 679; see Murzi and Steinberger 2017, 2). 'And' depends on us treating it as a conclusion of statements or treating it as a premise for conclusions. So commitment to A and B means being entitled to the conclusion A&B. This can count as an entry condition, while the exit condition means that commitment to A&B implies a commitment to the premises A and B. The problem is, as Fodor and Lepore point out, that the "expressions for conjunction (viz. '&' and 'and') appear on both sides of each equation" (Fodor Lepore 2007, 681). In other words, logical inferentialism is confronted with a problem of circularity (besides the question of the arbitrariness of rules).[71] As Fodor and Lepore argue, this problem can be circumvented by a practice-oriented approach. Accordingly, we would not have to make the rules explicit but accept them as an established fact. However, this approach gives rise to distinct challenges, such as the difficulty in distinguishing between coincidental adherence to a rule and intentional adherence (see Fodor and Lepore 2007, 683). In this regard, Fodor and Lepore argue that without a clear rule the inferentialist approach is useless (see Fodor and Lepore 2007, 684). Fodor and Lepore contend that logical inferentialism, due to its fundamental problems, is ultimately untenable and thus dismiss it. While delving further into these specific issues is not essential for understanding Brandom's approach, it is worth noting that logical inferentialism, despite its application in areas like natural-style deduction, remains a contentious theory.

71 While I will not delve into further details here, it is worth noting that informed readers will be aware of the challenges associated with the tonk-rule. It needs further arguments for why this objection can be disregarded or ignored.

Brandom effectively circumvents the challenges inherent in logical inferentialism by advocating that it emerges as a derivative of the normative language game. His objective is to employ a materialist inferentialist strategy to comprehend empirical concepts like 'tree', 'planet', or 'electrons'. Within this framework, speakers naturally recognize the dependence of utterances on specific purposes, anchoring logical reasoning in the human life-form. Notably, material inferentialism finds its most compelling illustration in Brandom's analogy involving "parrots, barometers, and thermometers" (Murzi, Steinberger 2017, 21). These objects causally interact with "stimuli" but do not 'respond' by using "concepts" (see Murzi, Steinberger 2017, 21). Conversely, full conceptual intelligence is evident in human assertions such as "this is red and it is a color" or "this is a vertebrate and it is a mammal" (Murzi, Steinberger 2017, 21). Logical inferentialism fails to support such assertions as logical operations are independent of specific content. For example, there is no logical connection between red to color. In contrast, material inferentialism explores the more universal terms and their intricate interconnections. According to this approach, assertions are intertwined with a social performative aspect known as "responsiveness to reasons" (Radford, 2017, 495).[72] This implies that an assertion acquires conceptual significance only when it can serve as a premise or a conclusion within a practical language game. The specific content of these language games is contingent upon the concrete human communities engaged in reasoning. In essence, content does not exist in isolation; rather, it is contingent upon the web of inferences within a particular human community.

Material inferentialism also comes with a variety of problems. One of them is the question of how children acquire the fundamental inferentialist tools and their normative habits. Fodor and Lepore remark that Brandom "is forever attributing to speakers/hearers commitment to norms, communicative intentions, interpretations" etc. (Fodor and Lepore 2007, 684). Although this is, indeed, an important question that I will discuss with regard to Habermas' criticism of Brandom, for now, I am more interested in how Brandom addresses the problem of mind-external reality. I will therefore proceed with explaining the inferentialist structure that is necessary for his proof.

As the meaning of an expression, according to inferentialism, is tied to the role it has in inferences, we can easily see that inferentialism must be holistic. When meaning is always produced through other meanings, the question emerges of where is the ultimate endpoint for constituting meaning. Brandom's

---

72 Radford, Luis. "On Inferentialism." Mathematics Education Research Journal 29, 2017: 493–508. https://doi.org/10.1007/s13394-017-0225-3.

inferentialism is therefore concerned with the semantic net, i.e., the whole of the inferentially related concepts. Meaning holism, in contrast to *atomism*, assumes that the meaning of all words is interconnected. There are therefore no stable reference points. Instead, we point out structural and partial relations between nodes within a semantic net. This meaning holism is compatible with forms of *molecularism* according to which words are bound to small groups of meaning. 'Kill', for example, would be bound to 'die' and 'cause' (see Jackman 2017). Each word is composed of smaller parts that, however, only in composition can provide meaning. A 'pandemic' (Murzi and Steinberger 2017) meaning holism goes beyond the limited meaning of molecules and atoms. We can find the first traces of meaning holism in analytic philosophy with regard to Quine's famous rejection of the distinction between analytic and synthetic judgments which brings him to the following conclusions: "It is misleading to speak of the empirical content of an individual statement" (Quine 1951, 40),[73] and that "the unit of empirical significance is the whole of science" (Quine 1951, 39). In these two statements, Quine expresses that there is no simple reference to an individual statement, but that it is the whole of science that makes the statement meaningful. Hempel comes up with a similar idea: "the cognitive meaning of a statement in an empiricist language is reflected in the totality of its logical relationships to all other statements in that language" (Hempel 1950, 59).[74] In other words, each statement is bound to a systematic whole that is implicitly presupposed by the speakers. Meaning holism then means that each individual statement receives its meaning through language as a whole.

Brandom follows an inferentialist holism since the basic structure of inferentialism cannot be atomist, as exemplified in the following statement: "For the content of each concept is articulated by its inferential relations to other concepts" (Brandom 2009, 15). So content is understood by inferences, and since every concept is understood by inferences that direct us beyond the immediate, concepts are determined by their relationship to a larger whole of concepts. Brandom claims explicitly that "inferentialist semantics is resolutely holist" (Brandom 2000a, 15). Knowing one concept requires knowing many since each concept is generated on the basis of other concepts (see Brandom 2000a, 15). Many, if not all concepts, are involved in the constitution

---

73 Quine, Willard Van Orman. "Main Trends in Recent Philosophy: Two Dogmas of Empiricism." *The Philosophical Review* 60 (1), 1951:20–43. http://www.jstor.org/stable/2181906.

74 Hempel, Carl Gustav. "Problems and Changes in the Empiricist Criterion of Meaning." *Revue Internationale de Philosophie* 4 (11), 1950: 41–63. http://www.jstor.org/stable/23932368.

of the content of one concept. However, if a multitude or even all concepts are involved, the issue of successful communication becomes problematic, as we can potentially refer to a single concept in numerous and potentially infinite ways. This constitutes the communication problem.

Inferentialism implies that speakers hold differently structured concepts based on the individual inferential networks through which humans confer meaning to concepts. This constitutes the communication problem that Fodor and Lepore express in the following question: "how can it mean the same to all of us if, on the one hand, its meaning is determined by its inferential role and, on the other hand, no two people could conceivably agree on all the inferences" (Fodor and Lepore 2007, 685). In other words, what does it mean to have shared concepts about objects in a community while each individual, however, has a different 'story' of inferences that constitutes meaning?

According to Mueller, Brandom defends a navigation model. It means that we do not share content, but give information on how to navigate around content.[75] There are some premises for understanding this. Most importantly, we communicate because we have different perceptions and conceptions of objects since we learn to know objects in individual settings. We integrate perceptual objects with regard to different inferences that we draw according to an individual, yet holistic, semantic net. In that regard, Mueller points out: "Brandom considers the systematic difference in information among individuals as the 'point' of communication" (see Mueller 2014, 141). So the communication problem is not a consequence of Brandom's inferentialism. Rather because we have the so-called communication problem, we need communication that Brandom attempts to explain with inferentialism. Because we have different 'experiences', we need to communicate and create a common language. This point of communication leads to the necessity of navigating "among perspectives without sharing contents" (Mueller 2014, 141). We do not have the correct content in our minds just because somebody utters a word that is associated with specific content in their mind. On the grounds of this, Mueller defines Brandom's meaning holism as the content of an expression being defined by the singular role the expression has in a complete system of conclusions to which the expression belongs (see Mueller 2014, 146). This system is the public and shared language of the communicators.

So since an expression can be interpreted only in dependence on the language used there is no single determinate content that is objective. There are

75 Mueller, Alex. "Löst Brandoms Inferentialismus bedeutungsholistische Kommunikationsprobleme?"*Zeitschrift für Semiotik* 34 (3–4), 2014: 141–185. https://philpapers.org/archive/MUELBI.pdf.

only users who attempt to agree on the content they discuss through their language. Content does not depend on the facts but on the method of how we attempt to make our different experiences meaningful to each other. Let us say, for example, I see an amphibious vehicle on a river and call it a boat, while another person sees it on the road and claims it is a car. In this case, both of our statements are interpretations depending on the contextual conditions through which we got to know the object. We both have reasons to call it this or that. If, however, any interpretation were possible, so that somebody else could call it a mule, another person could call it a planet, or another one calls it God, then we face the problem of contributing meaning to something at all. Obviously, it is reasonable to assume that something restricts our interpretative approach to the world. But how are we guided through a method so that we can reliably specify the meaning of our perceptual episodes with regard to assertions? We must somehow agree on a practice of how we construct objects. Reference, a gesture of pointing, is according to common pragmatist opinion not sufficient for such a specification. The common example here is Quine's 'Gavagai'. Referencing a perceptual episode of a rabbit by stating 'Gavagai' does not tell us what we ultimately mean or whether we actually point to the rabbit at all. For the amphibious vehicle, we somehow need to construct ways of connecting our different experiential episodes that lead to, for example, a gesture of pointing to the object in front of us. Another example is the question of whether a penguin is a bird. The example demonstrates that meaning is not a question of direct reference but a question of construction. If I say 'I see a bird', then 'bird' can mean a bird that is able to fly, or it might mean a bird that cannot fly such as penguins. People, indeed, get confused about such issues and discuss whether tomatoes are fruits or watermelons are vegetables. For birds, we usually assume that they can fly so that a penguin is not a prototype for being called a bird. Its classification depends on how we construct its meaning with regard to our semantic net. So when I say 'a penguin is a bird', I could refer to something that is capable of flying. In another instance, I might refer to something that is not capable of flying. The main point is that referring to something does not mean referring to its essence. We do not learn the meaning of 'bird' by merely referencing it nor would we learn the meaning of 'penguin'. In order to construct meaning, we need to, first of all, agree on a method of construction. In inferentialism, it means to draw inferences. Even referencing must be grounded on certain types of inferences. And this means that we are not talking about objects as merely existing in a mind-external reality that we simply point out, but it is first of all about how we agree on the construction according to which we integrate a perceptual episode in our semantic net. Whether a penguin is a bird or not, depends then on a material inferential structure of the semantic net according to which we reference something. In other words, we have to be clear about what we do when we use concepts. This brings us back to the pragmatic side of Brandom's project.

So after making us familiar with the basic semantic ideas of inferentialism, we can acknowledge that the meaning of a concept depends on what we do. So we attempt to understand semantics through the pragmatic side. A concept, according to inferential role semantics, is from a pragmatic standpoint a rule of how to act. A rule, moreover, is an instruction for a particular action (see Williamson 2010, 2). Dealing with concepts is therefore not simply a 'knowing that' like, for example, knowing the essence of a penguin and pointing at it. To know a concept does not mean mystically referencing something mind-external, i.e., connecting mind-internal objects with mind-external objects. Knowing is not about living in two different ontological realms that we connect by language. Rather, the concept gives implicit rules that we have to follow when we, for example, want to identify a particular object.[76] To know a concept is, therefore, to use a popular pragmatic catchphrase 'knowing how'. 'Real objects' are then not simply given in a mind-external world, but constructed with regard to the intended use of the speaker. For example, sitting in the passenger seat and asserting 'the traffic light is red' is a declarative sentence that, however, follows many implicit rules that are probably concerned with the dangers of the supposedly mind-external world around us, or with the perceived incompetence of the person in the driver seat. It is very likely not merely a true statement about the abstracted fact that there is a red traffic light. So one does not connect a mind-external object to a mind-internal object by language and then asserts a sentence so that everybody knows one did that. In this case, we should understand meaning more in terms of knowing how to practically navigate in a resistant or potentially even dangerous world. The assertion that the traffic light is red is then most likely a pragmatic response to the perceived danger of the particular situation of, for example, the driver being unfocused. As we can see, pragmatics offers an approach to meaning by integrating reference within the contexts of action and social communication.

We can enhance our understanding of the semantic net by either intuitively applying its rules within social interactions or by progressively clarifying the rules governing its usage. Furthermore, a word serves as a clearer expression of a conceptual rule within the semantic net. Through words, we can render

76 Fodor and Lepore can only point out ironically that an "introduction rule" for objects like 'tree' is hopefully "not some procedure for identifying trees as such since [... they are ...] a bit tired of verificationism" (2007, 680). This, however, is what Rockmore thinks Brandom's theory is (see Rockmore 2016, 112).

implicit concepts explicit by integrating them into our semantic network—our public language. We can even enhance their informational value by illuminating additional connections. For example, having an unknown mushroom in our garden is an implicit concept of this plantlike form. We hopefully all operate according to the implicit rule that an unspecified mushroom should not be eaten. If a mycologist (an expert on mushrooms) locates this mushroom in an explicit hierarchy of words, she can reliably add the rule 'not eating' this kind of mushroom (for example, false morels). In this sense, we make implicit rules explicit in the net of our language, and add information. We thereby store meaning in words and operate more reliably so that, for example, the unknown plantlike form can be now reliably categorized according to the edible and not-edible properties. In the public rule-book of our language, its property is then accessible by everyone through communication. As I understand it, a word is the explicit identification of a specific node within a semantic net. Conceptual rules can therefore be expressed, which means they are integrated into a public "rule-book" (see Williamson 2010, 2). It means, moreover, that the language learner does not produce all rules only for himself and keeps them in a private diary. Rules in the form of words are produced in a community explicitly and somehow stored in a language through a larger historical process.

There are many conceptual rules that we already follow implicitly. We only have to think about the difficulties we encounter if we are dating someone. It shows that not all of the rules are made explicit in our language, and many of these rules have not yet found a word that fits into our explicit language. However, implicit, unspoken rules also exist with regard to a community. Not following some of these rules, for example, can be sanctioned. At the moment rules are sanctioned, they are connected with a public force. In this case, they are not only some conventional rules but they are upgraded to norms, which usually involves making them explicit.

Now, what does this overall pragmatic understanding of language mean for Brandom's account? In the case of the resistant, supposedly mind-external reality, the members of a linguistic community charge extra-linguistic facts by identifying them with regard to their relevant set of rules. The idea is then that the mind-external world does not merely pop up in one's perception, but occurs with regard to the relevant rule-based activity of individual humans within their societies. Developing societies—if I understand Brandom right—convert more and more implicit rules into words according to the activity of giving and asking for reasons. The involved vocabulary then becomes more and more interconnected and thereby fine-grained with regard to, as Brandom argues, the mind-external world. So, for example, a stone-age society has produced less detailed theories about the 'atomic' deep structure of what is material. The nodes of their semantic net would have been wider apart regarding these contents and would have thereby allowed for more false explicit or false implicit connections. Modern society has tightened its semantic net with regard to atoms. It has come, as Brandom would claim, to a closer and more accurate view of mind-external reality. Moreover, as such a society also becomes more aware of its norms, it can sanction and regulate more beliefs. This expresses itself, for example, through social institutions that are allowed to punish. In this sense, our beliefs are regulated. For example, while it was more socially acceptable to believe the earth is flat in former societies, it comes nowadays with more disadvantages. Another example of the tighter semantic net is that nowadays, we entered the possibilities of harvesting nuclear energy on the basis of more detailed knowledge of the physical world. Unlike stone-age societies, we reconstruct our surroundings in terms of more sophisticated and socially interconnected words, in this case, specifically regarding the deeper atomic structure of objects.

According to this pragmatic theory, objects depend on the society and on the level of sophistication, which means that they constitute more rules that are explicitly stored in a larger net of a given public language. Now, the important point is that 'naked reality' does not simply justify our beliefs, but justification is a more complex social process that involves our rule-governed practices (see Huetter-Almerigi 2020, 3). For Brandom, there is no direct contact with the mind-external world so, for example, merely looking at mind-external reality does not teach us much.

So far it sounds simple and it seems to be common sense that we obviously improve our knowledge about mind-external reality by socially mastering concepts. However, the problem is that, according to this normative conception of meaning, the concept of mind-external reality is produced according to the norms of our normative community. It is, as I believe, an often unnoticed circle. This, as I call it, inferentialist circle constitutes "one of the most serious conceptual challenges", i.e., to demonstrate how we relate to a world beyond our social construction (Chung I Lin 2008, 105).

Given this first, rather simplified theoretical overview of pragmatics related to Brandom's inferentialism, we must recognize that because of the inferentialist circle, Brandom cannot mean to defend a simple mind-external, metaphysical realism. Considering the dependence of the object on the norms of communication, we rather deal, as I understand it, with procedural realism, i.e., we make explicit the rules by which we produce a notion of reality. Procedural realism means that although we charge all contents normatively, there is somehow an exit from the inside of our inferential language possible. Brandom, in particular, wants to open a door to mind-external objects. He believes this door towards 'attitude-transcendence' is part of our communication:

> A semantically adequate notion of correct inference must generate an acceptable notion of conceptual content. But such a notion must fund the idea of *objective* truth conditions and so of *objectively* correct inferences. Such proprieties of judgment and inference outrun actual attitudes of taking or treating judgments and inferences as correct. They are determined by how things are, independently of how they are taken to be. Our cognitive attitudes must ultimately answer to these attitude-transcendent facts.
>
> BRANDOM 1998, 137

So "things", which are independent of us, somehow fundamentally determine our 'normative' access to what counts as "conceptual content" (Brandom 1998, 137). In other words, although it is true that Brandom "dismisses the idea of semantic relations between words and elements of extra-linguistic reality" (McDowell 1997, 158), he believes that correct cooperation always involves an implicit, pragmatic understanding of a mind-external reality. This mind-external reality is somehow independent of norms and guides our discourse.[77] The argument itself is transcendental since for 'correct' communication there is a necessity for "attitude-transcendent facts" meaning 'real objects' which are entirely independent of us and which form an objective, mind-independent 'foundation' (see Brandom 1998, 137). They are the condition for the possibility of communication. Brandom plans to reconstruct the norms relating to these attitude-transcendent objects as guiding our practice of exchanging reasons. Thereby, he must show that the concept of reality that he presents is, indeed, outside of the inferentialist circle, which means that it is not convoluted with our own norms of communication.

Brandom's advanced strategy is certainly not a crude metaphysical realism that merely presupposes knowledge of mind-external reality. As I see it, it is a traditional form of argument that reminds us of German Idealism and in particular of Kant and Hegel. While Hegel, however, does not deny reality, he doubts similar epistemological (Kantian) efforts to distinguish what is constructed and what is mind-external reality. In other words, he doubts the distinction between form and content, which means that he is not only critical

77 McDowell, John: "Brandom on Representation and Inference." *Philosophy and Phenomenological Research* 57 (1), 1997:157–162.

of metaphysical but also of procedural realism. The distinction between form and content is difficult, if not impossible to draw. Brandom's intended procedural realism is, on the contrary, directed towards a mind-external reality (content) that is the point of convergence regarding our discourses (form). He believes that we can demonstrate how our discourse (form) is not only rooted in the relative normative application of concepts but ultimately grounded in this non-normative mind-external goal of discourse (the content). Although we could therefore say that Brandom reduces the 'object' of a mind-external reality by virtue of his inferentialism to a minimum (a transcendental remainder), the major weight of his theory still rests on this minimal, supposedly non-normative concept. It thereby is supposed to protect conceptual realism from turning into relativism.

I believe that Brandom's proposed pragmatic inferentialism without his proof is a viable strategy to discuss meaning. However, without this proof, it would, indeed, only suffice to create a relativist theory of meaning. Our relationship with mind-external reality is therefore, if I understand Brandom correctly, the 'absolute' condition for objective discourses. The major objection against Brandom is, however, that this mind-external reality as an objective condition is separated from our discourses and, as Huetter-Almerigi points out, understood as an unchangeable "bloc that stands in front of us" (Huetter-Almerigi 2020, 11).[78] I agree with Huetter-Almerigi, and this general criticism of Brandom as a mind-external realist builds the ground for my further interpretation.

Brandom's complex, linguistic model crosses semantics and pragmatics which leads to, indeed, very different interpretations of the theoretical aspect of mind-external reality. Because Brandom, on the one hand, anchors semantics in the pragmatics of inferential roles, but, on the other hand, anchors his pragmatics—as I will show further below in more detail—in propositional semantics, there emerges a debate whether Brandom is a realist or an anti-realist. In one of his first interpretations, Rorty, for example, identifies Brandom's approach even as open to the charge of relativism (see Rorty 1997b, 173–176). Other philosophers, however, celebrate Brandom for his mind-external realism that finally limits the open social game of language and ties it to a specific understanding of mind-external reality. Bransen can be seen as such an example:

78 Huetter-Almerigi proposes an interesting alternative of us being involved in the world so that we cannot divide form and content. I regard this to be the more Hegelian position.

> Given the open-ended nature of the game, it need not come as a surprise that experts working in the philosophy of science or in the sociology of knowledge who tend to be a bit on the pessimistic side might feel themselves inclined to endorse the thesis that objectivity will simply be unavailable in the final analysis. They might join forces with cheerful constructivists who embrace the thought that entitlements have in the final analysis nothing to do with objective conceptual contents, but merely with social forces that could be systematically structured and that are hopefully on their side.
>
> BRANSEN 2002, 389[79]

According to Bransen, Brandom finally solves the problem of anti-realist discourses by reconstructing how our communication is anchored in mind-external reality. As Bransen argues further, the "cheerful constructivists", or "these not so sophisticated anti-realists won't find Robert Brandom on their side" (Bransen 2002, 389). On the contrary, Bransen knows that "the expert who becomes familiar with the deep problems about knowledge, reality, objectivity, and truth" will agree with Brandom (Bransen 2002, 389). According to him, "the expert" needs to reconstruct Brandom's proof according to the following steps in order to understand discourses as rooted in reality:

1 We have to demonstrate first "that normativity cannot do without entitlements" to assertions (Bransen 2002, 389). Only if we are entitled to our assertions can here be a normative force. Entitlements are based on endorsements that are expressed as normative commitments in an interactive space of reasons (I will explain this condition in more detail below).
2 Structures of entitlements have ontological implications because, without these implications, we would not acknowledge the possibility of being correct. In other words, our interactions in the space of reasons are only possible on the grounds of reality.
3 The normative game of language will let the attitude-transcendent facts speak, and they will intervene in mistaken discourses.

Bransen delivers a helpful reconstruction of Brandom's realism. For now, I will point out, however, that all these claims are at least problematic. 1) The first claim assumes that entitlements as warrants for our assertions are the core of our communication. It thereby understands communication in its

79 Bransen, Jan. "Normativity as the Key to Objectivity—An exploration of Robert Brandom's Articulating Reasons." *Inquiry* 45 (3), 2002: 373–392. https://doi.org/10.1080/00201740276 0258204.

instrumental function. Language is used in order to achieve, justify or secure entitlements (this is also related to Brandom's claim that the assertion is the downtown of language, which I will discuss in point 2.2.2). Overall, this reveals Brandom's strong formal semantic presupposition. It is unclear, however, whether language really works in this formal way. According to a more Heideggerian understanding, we could also point to the discovering function that assertions have. Nevertheless, Brandom's semantic presupposition for the function of language is plausible, at least, insofar as it addresses the task of coordinating our beliefs in language. In other words, his approach would be, in my mind, successful if he suggested only a communication model. 2) The need for ontologies and the presupposition of an ontology on the basis of this need does not guarantee that ontologies are adequate. 3) It is questionable whether 'attitude-transcendent facts' utter themselves, or whether this is an anthropomorphism. In my view, facts do not act. This assumption needs more theoretical justification and will be also discussed in points 2.2.2 and point 2.2.3. Overall, Bransen's conclusion expresses a problematic excitement for a complex and controversial proof. Regarding our improvement with regard to approximating mind-external reality, Bransen proclaims further optimistically:

> appreciating these problems is a matter of playing the same game as the child whose cognitive development displays a rapidly growing success in coping with objective reality, namely the game of human understanding, the game of giving and asking for reasons—a game in which it is possible to become better.
>
> BRANSEN 2002, 389–390

I have discussed the problem of coherence with regard to Sellars' picturing before. I attempted to show that more detailed knowledge in one domain of knowledge depends on the conditions of our conception of knowledge in the first place, and so far does not secure a progressive approximation to a transcending mind-external reality. Rather, it is a progressive approximation toward a mind-internal standard. As I understand Brandom, he wants to show that this mind-internal standard is actually mind-external. For this, Brandom provides a new, creative, and more complex communication model than Sellars. So the question is: can Brandom's inferential role semantics anchor Sellars' inferential coherentism in a mind-independent reality? Bransen is very optimistic, but many other thinkers doubt that Brandom is successful. Margolis, for example, is critical of Brandom's "pragmatic extension" of the traditional, formal semantics. He particularly points to the necessary, but difficult idealizations that such a formal semantics has to undertake in order to reconstruct our complex, ordinary language: "notably, an analysis of the holist, contextual, societally embedded, historicized, tacit, profoundly informal, improvisational, consensually tolerated, and practically effective forms of discourse" (Margolis 2009, 184).[80] Given these complexities of our discourses, Margolis is skeptical of Brandom's "narrowly tailored tactics" (Margolis 2009, 185). He contends that Brandom's formal pragmatic semantics might be only relevant and possibly convincing to "the interests of a stubborn cohort of skillful specialists who are committed nevertheless to a project that was probably impossible" (Margolis 2009, 184).[81] Margolis implies that philosophers who are already realists will be convinced, whereas anti-realists will probably raise their eyebrows. The inner circle of the analytic-pragmatic specialists is interested in a transition from formal semantics to pragmatics, while Brandom's "brilliant" (McDowell 1997, 157) 'programmatic' approach 'promises' a linkage to mind-external reality, but, as Margolis judges, has "not yet delivered" (Margolis, 2009, 186). My intent is to show that beyond approaching problems of the particular pragmatic-analytic tradition, Brandom's account does not answer the fundamental question of how mind-external reality determines our discourses.

So Brandom adopts Sellars' inferential coherentism and extends it with an inferentialist proof. By adopting Sellars' inferential coherentism, however, Brandom also inherits the Sellarsian problems of relativist idealism. Sellars attempted to constitute objectivity in a coherent structure of inferences by applying the concept of picturing mind-external reality. Now, Brandom follows Sellars' coherentist, inferential strategy, but since he is more on the Sellarsian left wing, he distances himself from the idea of picturing (as explained before in point 2.1.3). According to the left-wing position, 'picturing' occurs as an external amendment to an actually inferentialist theory. Instead, Brandom wants to develop a concept of mind-external reality from the inside of inferentialism. In that regard, it is an interesting fact that the term '*experience*' does not appear in the index of *Making It Explicit*. The term may suggest a non-inferential contact with reality so that Brandom avoids it: "'experience' is not one my words" (Brandom 2002, 205). As Levine remarks, it possibly "gives rise to the Myth of the Given" (Levine 2012, 129), which means that there are "non-conceptual

80 Margolis, Joseph. "A 'Pragmatist' among Disputed Pragmatists: Robert Brandom's Between Saying and Doing: Towards an Analytic Pragmatism." *In Contemporary Pragmatism* 6 (1), 2009: 185–196.

81 Given the dismissive review of Fodor and Lepore, for example, Brandom's response gives testimony of his will to defend the issue and not engage in trench warfare. Unemotionally and not stubborn, he responds and thereby further elaborates the overall structure of his theory (see Brandom's response to Fodor and Lepore in *Reading Brandom* 2010).

episodes of immediate awareness that have epistemic *authority* simply through being given" (Levine 2012, 133). Since Brandom argues along the lines of inferentialism, authority, however, is conferred only by humans and not by facts or non-inferential episodes. In this sense, there can be no 'naked' experience as much as there is no 'naked reality' available. Without the concept of experience, however, the major question emerges of how to qualify one coherent system to be better than another one. Brandom claims that he has found a 'lever of objectivity' inside of the inferentialist paradigm that works without the 'unspecific notion' of experience. The idea is that individual terms, usually the non-conceptual content of statements, are normatively charged and understood through navigational, expressive actions of the speakers. They are not understood through deictic expressions just pointing out a 'mystical' connection between word and world. For example, saying 'this, here, now'—as the smallest unit of meaning in reference-based approaches—cannot fully constitute the seemingly simplest meaning of a present 'experienced' object. This means that when we refer to an individual appearance of color by saying 'this, here, now', we do not simply locate the individual color in a physical, present space-time and so determine its meaning. Rather, we use more complex concepts as navigational tools in order to locate the color event in our linguistic framework. These navigational tools start on the level of sentences that are embedded in inferences. In order to tackle this problem of deixis, Brandom develops the idea of anaphoric chains. Anaphoras are linguistic means used by speakers to refer back to a preceding subsentential term within a sentence or argument. For example, the word 'he' can be used as an anaphora: 'John is 20 years old and *he* is tall.' In Brandom's approach, anaphora is an intralinguistic tool that is supposed to explain how communicational success is possible without referencing individuals. Since anaphora penetrate, moreover, the whole linguistic net they connect the whole and contribute to its unity. In its central function, the concept 'anaphora' is probably one of the meanings in *Making 'It' Explicit,* while 'it' is the anaphoric connector. In other words, anaphoras are central to anchoring our words in the whole of our language. They serve as the conceptual glue to hold Brandom's system of inferentially related assertions together.

Brandom introduces anaphora as a crucial component of deictic expressions, as the latter alone are insufficient to adequately connect the inferential network of language to language entry events. The example of Nickel exemplifies the strengths of Brandom's anaphoric model best:

(A-i) Jane is tall. (A-ii) She is young. ∴ (A-iii) Somebody is young and tall.

(D-i) [pointing at Jane] She is tall. (D-ii) [pointing at Jane] She is young. ∴ (D-iii) Somebody is young and tall (Nickel 2012, 19).[82]

In (D), it becomes evident that deictic references lack the ability to establish connections between deictic claims. Each signification exists as an isolated moment, resulting in an inability to support the conclusive claims. On the contrary, anaphoras (A) offer a structured transition, enabling the possibility of syllogistic reasoning. Through anaphoric chains, contents are preserved and can be seamlessly integrated into the speaker's broader semantic network. Anaphoras allow us to trace every concept back to its originating event, thus facilitating the narrowing down of meanings within our discourse. By substituting an anaphora for the initial event, we progressively navigate through someone's arguments, reducing the range of potential interpretations. This ability stems from the fact that our language is primarily structured by inferential coordination rather than a direct representation of original events. In other words, without anaphoras, overall meaning would be absent, leaving us with mere fleeting instances of indicating something unrepeatable.

The concept of anaphora plays a crucial role in Brandom's later proof of objectivity. However, before delving into this proof, it is essential to provide further elaboration on Brandom's normative pragmatics. At the heart of Brandom's approach is the idea that knowledge is grounded in norms, commonly referred to as "norms all the way down" (Brandom 1998, 44). This implies that knowledge of expressions is contingent upon the rules governing our social contexts. Within this social context, Brandom focuses on the notion of the "objectivity of concepts" (Brandom 1997a, 156). It is worth noting that Brandom's work does not seek to justify individual statement contents as inherently true. Jackman (2017) discusses this Brandomian matter, stating that correctness is ultimately explained in terms of being correctly perceived as correct.[83] In other words, truth is understood through the practice of social discourse and expressed based on assertibility conditions rather than external truth conditions. The truth value of a sentence, therefore, depends on the norms employed in our social language games that enable assertions. Brandom suggests that these norms can be formalized, with a key norm being the assumption of responsibility for our own assertions. By making a statement, we publicly commit

82 Nickel, Bernhard. "Dynamics, Brandom-Style." *Philosophical Studies* 162 (2), 2012: 333–354, https://dash.harvard.edu/handle/1/13041028.

83 Jackman, Henry. "Meaning Holism." *The Stanford Encyclopedia of Philosophy* (2017). Edited by Edward N. Zalta. Accessed on November 25, 2017.

ourselves to it within our linguistic community, acknowledging the potential consequences of our words. Additionally, an assertion also involves claiming entitlement to that assertion. In this sense, we express (rather than merely state) mind-external facts from our perspective within the specific linguistic framework that we maintain individually and collectively. The justification for our commitments thus relies on the overall structure of this shared linguistic framework.

The crucial aspect for Brandom is that his inferentialism relies on securing an attitude-transcendent relation to a mind-external reality within our discourse community. In discussing this attitude-transcendent reality, we not only consider our normative statuses but also evaluate how our commitments correspond to mind-external objects. Through the process of sanctioning and replacing concepts that do not meet this mind-external standard, we gradually approximate reality. In my view, the approximation towards mind-external reality entails correcting our concepts by refining our linguistic net. This refinement is achieved by drawing more and more inferences. However, it remains uncertain whether Brandom's objective and correct procedural conditions for approaching mind-external reality can be deduced without explicitly defining the standard of mind-external reality. As previously mentioned, Brandom must avoid an essentialist ontology due to his commitment to inferentialism and the limitations of his normative account. Ultimately, it is a procedural approach rather than a straightforward metaphysical realism. Thus, the question arises: Can a mere procedural realism suffice? In this regard, Rockmore, as probably Brandom's greatest critic, finds the arguments for approaching mind-external reality unconvincing (Rockmore 2016, 112). He remarks that Brandom's "verification" of concepts is "no more than a promissory note" (Rockmore 2016, 112):

> In Brandom's position, verification consists in ascertaining which concepts are true through confronting them to reality. This is no more than a promissory note, which cannot be redeemed. The insuperable difficulty lies in going from fact-dependence to objectively correct inferences through relevant conceptions of normativity. That an inference is objectively correct according to prevailing normative standards is unrelated to what is really correct about the real. In short, the suggestion that inferences respect the facts according to the best current standards is insufficient to show they succeed in grasping how things are. In other words, as Kant points out in

> his Copernican turn, no one has ever been able to show how one can correctly infer from appearance to reality.
>
> ROCKMORE 2016, 112[84]

Rockmore argues that obtaining objective inferences through normativity by asserting a fact-dependence is impossible. He contends that our criterion for determining objective correctness is detached from what is real. However, I do not entirely concur with Rockmore's argument that comprehending mind-external reality is impossible. In fact, I am uncertain whether it can be classified as possible or impossible. This uncertainty arises because relativism, in its essence, cannot definitively establish the impossibility of something without contradiction. Relativism inherently maintains an open stance, acknowledging the potential for the conceptual or alternative means of revealing mind-external reality. Nevertheless, I do agree with Rockmore's observation that Brandom fails to demonstrate how a mind-external reality influences our discourses. To substantiate this assertion, a further, closer examination of Brandom's inferentialist conceptual realism is necessary.

I have outlined the main objectives and provided a simplified overview of *Making It Explicit*. It is evident that I interpret Brandom as having implicit inclinations towards mind-external realism, despite his explicit intention to avoid it. He situates his project firmly within the domain of an inferentialist, normative pragmatics, which I construed as a form of procedural realism. In the next point, I will delve deeper into the foundations of his procedural realism, specifically by examining the internal structure of language with a focus on assertions. Once this groundwork is established, I can finally proceed to discuss his proofs of objectivity in point 2.2.3.

2.2 *Procedural Realism and Assertion*

With Brandom's complex and sophisticated theory comes a practical, hermeneutic problem. Firstly, readers must familiarize themselves with the extensive terminology borrowed from the broad analytic tradition, as well as the numerous new concepts he introduces. Moreover, grappling with roughly 550 pages of dense philosophical arguments is necessary before his project culminates in a series of arguments for the objectivity of his inferentialism. Hence, the question remains whether my necessarily condensed and simplified reconstruction adequately captures his intentions and methodology.

---

84 Rockmore, Tom. "On the Pittsburgh School, Kant, Hegel, and Realism." *Wilfrid Sellars, Idealism, and Realism: Understanding Psychological Nominalism.* Edited by Patrick J. Reider. London: Bloomsbury Publishing, 2016.

Moreover, interpretations of Brandom's project can vary, with some perceiving him as a realist and others as an anti-realist. For instance, Rorty, in his initial reading of Brandom, characterizes his inferentialism as "wide-open to the charge of 'relativism'" (Rorty 1997b, 173, 176). In that regard, he also states that Brandom's account leaves "no room" for a traditional "unconditional":

> There is, as far as I can see, no room for anything unconditional in Brandom's view of things, nor any room for a 'foundation' for an existing consensus. I take it that for Brandom no principle can claim a higher status than being presupposed by some 'de facto' established practice.
>
> RORTY 1997b, 176[85]

Moreover, going beyond the rejection of an unconditional stance, Rorty, being intimately acquainted with Brandom's work as his doctor father and as a close friend, even goes as far as to argue against the notion that Brandom's inferentialism aligns with a concept of progress towards reality:

> The intuition that we are making intellectual progress is simply the intuition that, in respect to self-consciousness and intellectual responsibility, we are getting farther and farther away from the cavemen; it does not need backup from notions like 'closer to Reality' or 'more nearly universally valid'. This would be analogous to saying that the intuition that inquiry is in touch with reality is simply the intuition that it is constrained by reality; it does not need backup from notions like 'corresponding' or 'mapping'.
>
> RORTY 1997b, 176[86]

As evident, Rorty rejects the notion that we can approach mind-external reality. Consequently, we can classify Rorty as anti-Sellarsian, despite Rorty's support for inferentialism, as Sellars' central contention is that we can progressively approach mind-external reality. It therefore becomes particularly puzzling when Rorty states that it is hard to find "any interesting point of difference" between Brandom and his own views (Rorty 1997b, 173).

From my perspective, Brandom aligns himself with the Sellarsian position, which implies his commitment to what I perceive as a form of mind-external

85 Rorty, Richard. "What Do You Do When They Call You a 'Relativist'?" *Philosophy and Phenomenological Research* 57 (1): 173–177, 1997. Accessed on September 04, 2017. http://www.jstor.org/stable/2953787.

86 Rorty further concludes that philosophy should rather ecumenize than revolutionize (see Rorty 1997b, 176).

realism and, consequently, an anti-relativist stance. However, whether this stance leans more towards an amended mind-external realism or a procedural realism ultimately hinges on the effectiveness of Brandom's strategy for providing a proof of objectivity. If Brandom is successful, it is probably a "knock-down proof" for many opposing relativist theories including my own (Chung I-Lin 2008, 109).

Interestingly, Brandom's chosen inferentialist-pragmatic strategy to defeat relativism is usually associated with its vulnerability to relativism. If Brandom's inferentialism does not want to leave the "door open for foundational skepticism", he, indeed, needs to show how our normative practices are "suitably *constrained* by how the world actually is" (see Reinikainen 2021, 56).[87] Huetter-Almerigi observes regarding this challenge that Brandom operates in "two realms that need to be bridged" (2020, 6). Brandom does not only move from semantics to pragmatics, and from pragmatics back to semantics, but he also moves "back and forth" between representationalism and inferentialism. That is because he wants to prove that "facts [...] supersede vocabularies" even if they are "domesticated" by vocabulary (Huetter-Almerigi 2020, 6). So although inferentialism is opposed to the "mind-world split" (Radford 2017, 495) that usually comes with representationalism, Huetter-Almerigi thinks that his approach to secure objectivity re-invites old representationalist tropes of a stable, unchangeable world. His account is thereby open "to the accusation [...] of the old scheme-content distinction", i.e., the distinction between content and form that Hegel, for example, rejected (Huetter-Almerigi 2020, 6).

Similarly, just as inferentialism is commonly linked to anti-realism (see Veluwenkamp 2021), normative pragmatics is often perceived as being in opposition to realism. Grönert, for example, locates the conflict of objectivity not majorly on the side of Brandom's inferentialism, but on the side of his normative pragmatics. On the one hand, Brandom argues that there must be an objective norm by virtue of which language users can reliably identify their mistakes and approximate mind-external reality. On the other hand, norms are dependent on the attitudes of the language users (see Grönert 2005, 161).[88] In other words, if we cannot exclude that we can all be collectively, and objectively mistaken, then there is no exit towards reality possible. Brandom calls this condition the '*No Communal Error*' which is an undesirable "consequence of *I-we* theories of conceptual norms" (Brandom 1998, 603). It means that normative

87 Reinikainen, Jaako. "Brandom and the Pragmatic Quest for Semantic Objectivity." *Acta Philosophica Fennica* 97, 2021: 55–78. https://philarchive.org/archive/REIBAT-4v1.

88 Grönert, Peter. "Brandom's solution to the objectivity problem." *Pragmatics & Cognition* 13 (1), 2005: 161–175. https://doi.org/10.1075/pc.13.1.12gro.

pragmatics based on a shared net of concepts may lock communities inside of relativism. Somehow his approach, therefore, needs to 'triangulate' the relationship of the norms that communities impose on subjects. In so far as Brandom wants to make his inferentialism work, this "*objectivity problem*" is then "the central challenge" of *Making It Explicit* (see Grönert 2005, 162).

The first intuition is now that we need a bird's-eye-view. Brandom, however, denies this as a genuine possibility: "there is no bird's-eye view above the fray of competing claims from which those that deserve to prevail can be identified, nor from which even necessary and sufficient conditions for such deserts can be formulated" (Brandom 1998, 601). For gaining an objective, stable ground, Brandom wants to extract an objective, attitude-transcendent super-norm from the inside of our played language games. His solution needs to focus on the speakers who transcend their norms. More precisely, the speakers must transcend their perspectives when they play language games of giving and asking for reasons. It is not about grasping mind-external reality directly, but about demonstrating how they follow a norm of improvement that ultimately is guided towards mind-external reality.

So Brandom's intentions, as I understand him, are genuinely proceduralist since he wants to prove objectivity only with the self-forged tools of inferentialist semantics from the inside of reflective, normative pragmatics. His promise is to extract a concept that makes sentences true from the inside of our existing linguistic system (see Fodor and Lepore 2001, 480).[89] But how successful is he? In my opinion, his intended, sophisticated, complex procedural approach conceals the mind-external realist ideas that he introduces from the outside. The final proof reminds one a bit of the German Idealist tradition since he applies their old trick of reflecting on the necessary conditions for our knowledge (see Brandom 1998, 601–602). In Brandom's case, it is also not the philosopher-king who stands on the sidelines of discourses and who extracts the pure theory of knowledge from a neutral observing position—a standpoint from nowhere. Having learned the Hegelian lessons of a more democratic history of practically involved, finite human beings, it is the speakers themselves who ultimately must transcend their perspectives by a self-reflective practice within their language games. They do this not with a theoretical attitude but quite practically. While they communicate, they constantly produce the norm of objectivity that—if Brandom is right—can lead them to increasingly better knowledge regarding mind-external reality. Their 'attitude-transcending

89 Fodor, Jerry, and Ernie Lepore. "Brandom's Burdens: Compositionality and Inferentialism." *Philosophy and Phenomenological Research* 63 (2), 2001: 465–481. doi:10.1111/j.1933-1592.2001.tb00119.x.

attitude' is related to the structure of giving and asking for reasons and more specifically to the way they treat assertions with regard to the contained propositions.

Regarding the establishment of a shared understanding among speakers, Brandom asserts that our rational discourse is intricately linked to the act of assertion and its necessary propositional structure. According to Brandom, "asserting is the fundamental speech act" (Brandom 1998, 173). Through its propositional structure, assertions can act as crucial connectors in syllogistic reasoning (see Brandom 1998, 167). Not only does the proposition serve as an inferential link connecting assertions to other assertions, but it also plays a crucial role in bridging the semantic and pragmatic aspects of Brandom's philosophy. On the semantic side, assertions are treated as propositions, and they are often reduced to their propositional content. However, Brandom swiftly dismisses Frege's definition of assertion as: "the declarative sentence *p* as 'an undertaking to the effect that *p*'" (Brandom 1998, 173). As a thinker influenced by the pragmatic turn, Brandom finds this assumption about assertions to be "disappointing" (Brandom 1998, 173). It fails to capture the essence of what speakers actually do when they assert. For instance, if a student enters a lecture and asserts 'earth is flat,' it would undoubtedly have a range of consequences. However, the true absurdity of this situation does not solely lie in the statement *p* itself. What makes it truly absurd is the fact that it is highly unlikely for someone to enter a public space and randomly make such an assertion without any intention or relevance. Similarly, if the student were to assert 'the earth is a sphere,' it would be equally absurd. Here, we can observe how the act of 'merely' asserting p fails to align with our norms of behavior in a classroom. The student may be engaging in the language game of a naked assertion, but the situation does not permit this specific Fregean speech act. As Austin would likely argue, there is no clear illocutionary force in the student's speech act, resulting in random perlocutions as its consequences. Thus, while there is propositional content in the assertion, we must understand this content from a pragmatic perspective, considering how assertions are charged by commitments on the pragmatic side.

The evident absurdity of randomly occurring assertions that declare arbitrary propositions aligns with the insights of the pragmatic turn. Speech acts of assertion trigger intricate chains of interpretations and reactions within the social group, which cannot be simply reduced to the mere effect of declaring a proposition *p*. The provided example encompasses numerous contextual conditions that shape the meaning of the student's speech act. Nevertheless, Brandom's focus does not lie on the many pragmatic conditions that underlie speech-acts, but lies primarily on the central, formal pragmatic condition of asserting something. When addressing the question of "what is it that we are doing when we assert [...]?" Brandom identifies one central pragmatic action and answers: "We are undertaking a certain kind of commitment" (Brandom 1998, 167).

The notion of commitment extends beyond the semantic understanding of an assertion as a mere belief. Brandom links the act of assertion to its normative dimension. It implies that when we utter an assertion within a linguistic community, it entails certain responsibilities (see Rosenkranz 2001, 232).[90] Through assertions, we enter the open "public arena" and make "sentence[s] available" to others (see Brandom 1998, 170). If they accept the assertions as a source of authority, they are 'licensed' to do things with them (see Brandom 1998, 170). But, on this basis, assertions can also be sanctioned "if the propriety of these commitments cannot be substantiated" (Rosenkranz 2001, 233). So assertions are also tied to what one ought to do and part of it is to play the propositional game the correct way. Now, Brandom has to explain and prove how the game is played the correct way.

Brandom further specifies that asserting involves three dimensions: first, committing to a given propositional statement; second, committing to its consequences; and third, claiming some form of entitlement for making the assertion. These dimensions connect us to our linguistic community. The first dimension, commitment to a proposition, is semantic, while the second dimension, taking responsibility for consequences, and the third dimension, implicitly claiming warranting entitlements, are pragmatic. In other words, assertions serve as the fundamental bricks for constructing our linguistic communities, facilitating the exchange of reasons on both pragmatic and semantic levels.

Brandom discusses now in inferentialist terms how the assertion can count as an "implicit knowledge claim" (Brandom 1998, 201). In contrast to representationalist approaches, which define truth conditions, inferentialism rejects theoretical access to presumed extra-linguistic truth and instead emphasizes the importance of assertability conditions. This raises the question of how classical knowledge, defined as true, justified belief, fits within the context of inferentialism (see Reinikainen 2021, 57). Firstly, justification in inferentialism implies that an assertion functions as the conclusion of an inference. In other words, the assertion must be supported by a chain of reasoning (Brandom 1998, 204). Secondly, when making an assertion, the speaker is claiming a

90 Rosenkranz, Sven. "Farewell to Objectivity: A Critique of Brandom." *The Philosophical Quarterly* 51 (203), 2001: 232–237. doi:10.1111/j.0031-8094.2001.00226.x.

"positive justificatory status," which Brandom identifies as the belief component (Brandom 1998, 204). However, a challenge arises when attempting to determine the truth of an assertion within the inferentialist framework. There is no straightforward argument or method of ultimate verification available. This limitation stems from the fact that every assertion relies on an "inheritance of entitlements" (Brandom 1998, 205). This leads to an inherent regress on the side of inferences, necessitating an endless demand for further justification (Brandom 1998, 205). Consequently, one may naturally question the nature of these entitlements if they lack ultimate grounding.

The issue of initial entitlements exposes the inherent circularity within an inferential system. The process of justification knows no end, as every seemingly foundational assertion is susceptible to challenge. None of the entitlements are "immune to criticism", as Brandom aptly notes (Brandom 1998, 204). But as Brandom also remarks, there must be "prima facie entitlements available to get the process off the ground" (Brandom 1998, 204). As a pragmatist, Brandom believes in practice: "One must start with a notion of taking or treating inferences as correct in practice" (Brandom 1998, 205), since "without such a practice, there is no game of giving and asking for reasons to bring inferences into in the form of explicit assertions" (Brandom 1998, 205). In essence, the acknowledgment of certain truths is imperative for effective communication to occur. Without this implicit recognition, our ability to engage in meaningful discourse would be rendered impossible.

Brandom pragmatically safeguards his approach by acknowledging that the challenge of entitlements also requires justification. As a consequence, "claims are treated as innocent until proven guilty" (Brandom 1998, 177). In this regard, Brandom also discusses Wittgenstein's insights regarding rule-based approaches, highlighting a similar problem of infinite regress. Any understanding based on rules encounters the challenge of an infinite chain of rules, as we require rules for applying rules and rules for applying those rules of rules, and so on. However, historically and socially, the regress has come to a halt through human action and the establishment of rules through practice. Brandom's solution to the infinite regress of rules acknowledges that rules do not emerge from an internal source of thought or a foundational philosopher who sets out fundamental rules. Instead, they have already been established by preexisting human communities with preexisting understandings. Brandom explains: "grasping a rule without interpreting it is grasping it in practice, rather than by substituting one expression of a rule for another" (Brandom 1998, 65). However, this emphasis of practice also brings us back to square one, the question of the relationship to truth and whether there is attitude-transcendence.

Examining the semantic-pragmatic model that arises from successful practice, Brandom argues that meaning is, indeed, dependent on the justificatory processes within society. While Rorty, with his emphasis on justification, abandons the concept of truth, Brandom, however, seeks to retain it. In this regard, he rejects *regularism*: the behaviorist or naturalist idea that we follow norms depending on the authority that establishes them. If norms were solely dependent on the prevailing authority of a society, they would be relative. Brandom therefore insists that there must exist "an *objective* sense of correctness that governs their [concepts] application—a sense of appropriateness that answers to the objects to which they are applied and to the world of facts comprising those objects" (Brandom 1998, 594). Assertions must be guided by an implicit normative rule that transcends their societal determination. Can we extract this rule? Brandom refers in this regard to Sellars' pragmatism of rules: "a rule is lived, not described" (according to Brandom 1998, 25). I think this claim remains central. So the solution cannot be found in a semantic extraction of 'a truth' on the theoretical side. But—and herein lies the key to Brandom's system—truth is related to action:

> We are believers, and believing is taking-true. We are agents, and acting is making-true. To be sapient is to have states such as belief, desire, and intention, which are propositionally contentful in the sense that the question can appropriately be raised under what circumstances what is believed, desired, or intended would be true. Understanding such a content is grasping the conditions that are necessary and sufficient for its truth.
>
> BRANDOM 1998, 277

First of all, we can see that truth is related to our actions of making something true. So we clearly shift into a proceduralist account. Second, if we want to make truths, we cannot simply trust the authorities. Rather, we align, and this is Brandom's solution, our mental states with propositions so that we generate conditions under which something could be wrong, independent of our community. Basically, it means that we act by investigating the truth of the statements of others and thereby enter a more objective lane that dissociates ourselves from the individual and the community perspective. Attaching to the propositional assertion is here the norm that guarantees that a statement could be wrong independent of our shared beliefs. This practice breaks then the structures of merely authoritarian entitlement, and allows us to, according to Brandom, objectively approximate mind-external reality. Although we will never reach a true concept, we eliminate what is wrong. In the end, the adoption of propositional language games contributes to a tighter semantic, propositional net.

So it comes down to Brandom's semantic imperative: 'You shall use propositional assertions!' Brandom wants to show how the imperative of adopting propositional assertions works through the score-keeping model. How this, however, plays out in detail will be discussed in the next point (2.2.3). For now, it is interesting that, on the one hand, Brandom refers to successful practice. Prima facie entitlements somehow depend on society and its institutions. Truth depends on their authority. On the other hand, Brandom focuses on the assertion and its propositional content. Here, he wants to break the authority of mere community-based commitments and introduce an objective lane of justification. We can accept this model on quite pragmatic grounds since truth is part of our doing, while, at the same time, we are skeptical of societal authority alone. Despite its plausibility, however, the model somehow also reveals a very formal focus with its final result: the negotiation of propositional assertions. The propositional assertion fulfills different formal requirements in Brandom's system. Its intended purpose is to serve as the formal framework through which we can approach mind-external reality, transcending individual perspectives. However, one must question whether humans are truly equipped to exchange propositional content solely through assertions or if there are additional facets at play. Does the assertion, with its propositional content, accurately capture the complexities of how real human beings communicate and create meaning? Here, it is reasonable to object whether its formalization inadvertently diminishes and oversimplifies human interaction. It is important to examine objections that challenge the notion that asserting alone is the essential activity in our human exchange of reasons.

At the beginning of *Making It Explicit*, Brandom states that he is concerned with the rather relativist anthropological question regarding the meaning of the word 'we'. Who is it that 'we' refer to when we say 'we'? However, unlike Hegel's approach, Brandom's work lacks a historical and sociological discussion of this aspect. It could be argued that this changes with *A Spirit of Trust*, but upon closer examination, it becomes evident that Brandom primarily focuses on describing how societies adhere to his concept of inferentialist pragmatics. Instead of engaging in a detailed sociological analysis, he formalizes the fundamental human activity of asserting to define this collective 'we'. Brandom, being a linguistic rationalist,[91] places less emphasis on the empirical

91 Hegel's philosophical position does not align with rationalism. Instead, he perceives himself as situated at the culmination of a dialectical progression that surpasses empiricism in favor of rationalism, and subsequently integrates rationalism and empiricism into critical philosophy. Finally, he further synthesizes the contradictions emerging from critical philosophy into his own objective idealism. The precise meaning of objective idealism

aspect, as he often highlights. His commitment to the assertion reveals a dual nature: pragmatic and semantic. According to Brandom's pragmatic framework, assertions are fundamentally governed by community rules. However, he rejects an absolute dependence on 'human' rules and upholds conceptual realism. Throughout his work, Brandom repeatedly assures us that our adherence to rules is a fundamental response to the way the world truly is. This embrace of conceptual realism diverges from the more pragmatic and pragmatist understanding of the intricate nature of human life. Instead, it leads us towards the formal concept of propositions. The conceptual realism, therefore, leads away from the rather pragmatic and pragmatist understanding of the complex life-form of humans. It leads us to the formal idea of propositions. Brandom posits that the essence of human beings lies not in their utterances but in their ability to assert, which enables a connection to truth. He asserts, "practitioners who can produce and consume assertions are linguistic beings. Practitioners who can produce and consume reasons are rational beings" (Brandom 1998, 204). However, this view does not align with the anthropological work undertaken in philosophy since Hegel. The reduction of humans to their capacity for reason, presented as assertion-making animals, contradicts the postmodern traditions that predominantly explore the empirical side of human existence and language. I can accept that Brandom is interested in human beings in so far as we are formally rational, but it opens the question if his method really discusses the essence of human beings. Brandom might counter that his approach concerns how we ought to act rather than providing an empirical description of our current state. However, this raises further questions about whether his formal imperative ('You shall use propositions') can be derived without considering how human beings truly live. In essence, this objection echoes the longstanding Hegelian criticism of Kantian formalism.

To be fair, Brandom does not say that only language or the assertion makes us human. In that regard, Brandom himself makes some relativist remarks. In his response to Taylor, for example, he argues for the possible existence of other forms of communication. There, he affirms the question of whether there could be "*rational* creatures who are not *logical* creatures" (Brandom 2010, 304). Brandom suggests the possibility of lifeforms that do not engage in discussions of reality using comparable models of inference. Moreover, he affirms the possible existence of creatures who do not rely on the "expressive

poses a challenging question, and I hold the contentious perspective that it represents a form of historicist relativism. In light of this interpretation, Hegel does not prioritize the assertion as a central element, as it would merely serve as a formal condition within his framework.

resources provided by propositional-attitude ascribing locutions" (Brandom 2010, 304). Brandom's perspective appears to be open to the possibility of other forms of reason that are not based solely on assertions. However, it is also true that Brandom draws a distinct line between animals and humans, displaying a somewhat dismissive stance towards animal nature. For example, it seems as if, according to Brandom's introduction of inferentialism, animals respond to room temperatures in a manner akin to thermometers. This position has raised concerns for Rorty, who questions whether Brandom is neglecting certain fundamental responsibilities of a philosopher: "saying that babies and dogs have beliefs only in a derivative and metaphorical sense [...] is to decline responsibility for explaining how they differ from thermostats" (Rorty 2021, 175). In light of this argument, certain aspects of Brandom's responses can indeed be called into question. For instance, when addressing Taylor's objection regarding the various forms through which humans engage with the world, such as art, poetry, or music, Brandom acknowledges them as "issues of absolutely vital importance to human life" (Brandom 2010, 303). Yet, in the same sentence, he declares that "after all, [we are] not only rational but animals" (Brandom 2010, 303). This raises the valid concern of whether Brandom overlooks a fundamental aspect of our humanity. Can we truly categorize art, poetry, and music as mere expressions of our animal nature? By conveniently sidestepping the question of what renders reason itself reasonable, Brandom seems to disregard the self-reflective task—asking who 'we' are—that holds significant value within the Continental tradition. Instead, Brandom adopts a rationalist perspective that confines his exploration within the boundaries of reason—'we' are reasonable beings—considering reason to be the essence of normativity. From within the realm of reason, he emphasizes the assertion as the key element of rationality.

Brandom utilizes another metaphor to expound on the centrality of assertions. In the same response to Taylor, he rebuts Wittgenstein's idea of language as a city without a downtown: "But I think language does have a downtown, and that it is the practice of giving and asking for reasons" (Brandom 2010, 301). This downtown is "essentially" the assertion with its propositional structure and nothing else "deserves to be called a 'language' unless some of its performances have the practical significance of claims about how things are" (Brandom 2010, 301). However, if communication can occur through various means, as Brandom acknowledges, the question arises as to why we should solely adhere to his propositional language at any given moment. Once we abandon the notion that there exists a single universally valid procedure, it appears that our language is based solely on our agreement to adopt a conventional procedure within our particular, historical community. This could involve embracing formal, Brandomian semantics as a linguistic framework, but it could also encompass engaging in dialectical conversations that lack an absolute foundation. From my perspective, such an agreement on how we communicate would be the relativistic outcome of our quest for human identity. Reason, in my characterization, becomes the inherent openness of human beings towards diverse modes of communication. In this regard, I am inclined to prefer Hegel's model, which delves into the exploration of humanity and its self-imposed obligations that have evolved throughout history.

So can we really say that language has a downtown consisting of propositions? I mean I agree that propositions are useful tools, but should they be the center of our reason? *Discourse Analysis*, an empirical branch of *General Linguistics* would have some objections. Gee, for example, states:

> In fact, I believe this is simply a prejudice on the part of academics who believe, often falsely, that what they themselves primarily do to and with each other is 'exchange information'. Language, in fact, serves a great many functions, and 'giving and getting information', yes, even in our new 'Information Age', is but one, and, by no means, the only one.
>
> GEE 1999, 1[92]

In empirical research, it is uncommon to consistently identify the content of communication solely through clear propositions embedded in performatives. Instead, we encounter the persistent reality that there is no distinct center of our communication. Language, as a complex phenomenon, intricately weaves together the various aspects of society, encompassing both micro and macro spheres. It encompasses social institutions and individual ways of life. These individuals express themselves through the arrangement of their surroundings, be it art, philosophy, poetry, rituals, bodies, politics, or music. Language also manifests in casual conversations, jokes, and even in smaller instances of aesthetic absurdity. Can we simply reduce this rich "social soup" (Brandom 1998, 54) to the imperative of engaging in a formal exchange of propositional reasons?

I understand that philosophers interested in morality pose distinct questions that are less focused on the empirical aspects of human life. They do not inquire about moral facts, but rather about what actions *ought* to be taken. Brandom, from a rationalist perspective, engages with a normative realm

---

92 Gee, James Paul. *An Introduction to Discourse Analysis—Theory and Method.* London: Routledge. 1999.

of reasons. In response to these objections, Brandom can maintain a strong reliance on a rational, non-empirical standpoint. In an aptly titled interview called "Inferential Man," he then defines philosophy as the exploration of what it means to be human (Williams 2013, 368). For Brandom, being human is defined "by living in a normative space of reasons" (Williams 2013, 368). However, can we adequately describe this realm of reasons solely through propositional assertions? Indeed, there is limited information available about Brandom's stance on recent developments in philosophy, such as feminism, minority rights, and critical race theory. These emerging disciplines challenge our 'rational' self-conception as normative beings, a point that many Continental philosophers would acknowledge. Through these disciplines, philosophy actively engages with the pragmatic and empirical aspects of human existence. It seems reasonable to consider that these disciplines can provide insights into the nature of our normative realm. Such positions are guided by a skeptical view towards a Kantian formalization of what we ought to do. Instead, they derive reflective oughts while also recognizing the empirical dimensions of human life.

I understand that this objection may appear somewhat unfair considering the extensive scope of Brandom's research and his explicit focus on formal rationality. Nonetheless, given the nature of Brandom's philosophy, which intends to explore the essence of human rationality, it seems reasonable to pose this question to him as a philosopher. It is worth noting that the University of Pittsburgh, known for its Hegelian school, often ranks highly in philosophy-related evaluations. This recognition is likely attributed to the profound impact and influence of its renowned philosophers within the discipline. However, one might question whether it is mere happenstance that Brandom, also referred to as the "philosopher's philosopher" (Gleeson 2020), remains relatively unknown outside of expert circles and even among scholars in his own field.[93] Notably, figures such as Chomsky, Žižek, Butler, West, Habermas, and Sloterdijk, among others, contribute to public discourse and enjoy broader recognition. Yet, Brandom, despite his heavyweight status in the realm of philosophy, remains somewhat unheard of within the wider philosophical community.

Recently, Brandom published his book *A Spirit of Trust*, aiming to incorporate Hegel's philosophy into an analytic discourse and render it relevant to our contemporary era—an undoubtedly commendable achievement on

93 Gleeson, Jules. "Robert Brandom, a Philosopher's Philosopher" *JstorDaily.* 2020. https://daily.jstor.org/robert-brandom-a-philosophers-philosopher/.

Brandom's part. However, it is crucial to recognize that Hegel's philosophy grappled with the practical challenges of its own historical context. In light of this, one might question whether Brandom's formalized approach adequately captures the essence of Hegel's philosophy in relation to our present moment, potentially overlooking the practical aspects that could also contribute to a comprehensive understanding of philosophy and ourselves.

Considering Brandom's focus on normative pragmatics and inferentialist semantics, it is understandable that he places the assertion at the core of his philosophical model. As the convergence point of semantics and pragmatics, the assertion plays a pivotal role in his framework. However, one might argue that Brandom's approach could benefit from a complementary element—a phenomenological reduction that systematically isolates the assertion from our cultural formations and concrete practices. Hegel's use of the *Phenomenology* as a methodological example leading up to the *Logic* exemplifies such an approach. By incorporating a similar reduction, Brandom's philosophy could potentially achieve a more comprehensive and nuanced perspective.

But I admit again that these objections may miss the point of Brandom's formal focus on philosophy. As Brandom understands himself "as an analytic philosopher", his training rather goes along with the analytic "strength" of 'demanding' "clarity" (Williams 2013, 370). So I may conflate here the public fame and public interest with his scientific significance in terms of advancing a discipline in its rigor and details. I may miss the "scientific control over an understanding of meanings" that comes with a rather fine-grained approach that cannot enthuse a possibly false audience (Williams 2013, 371). Moreover, there is an academic right to be silent and to focus on a form of expertise that does not have to be related to an increasingly troubling, public discourse. So my objections aside, I should finally focus on Brandom's proof of objectivity that may finally accomplish Sellars' original goal: to defeat ('bad') idealism.

2.3 *Proof of Objectivity—Towards Mind-External Reality*

In the climactic finale of his book *Making It Explicit,* Brandom finally presents his "knockdown" (Chung I-Lin 2008, 109) "master argument" (Kusch 2002, 253).[94] The wording seems ironic, but Brandom's use of extensive and

94 Since a successful argument defeating relativism is a desideratum of philosophy, such arguments attract the attention of many critics. In fact, I could only find a few who agreed that Brandom has achieved his goal of objectivity (an exception, for example, is Bransen 2002). While, for example, Chung I-Lin (2008) and Rosenkranz (2001) point at flaws of the formalized version of his proof. Kusch (2002), Prien (2010), and Reinikainen (2021) look at the meta-semantic arguments. They cannot see that Brandom's proof successfully

detailed terminology spanning around 500 pages to build his complex discourse framework, aimed at achieving objectivity, has created specific expectations. Moreover, Brandom asserts that he does not know "any other account" that tried "baking an objective cake out of social ingredients" (Brandom 1997b, 200–201).[95] The master idea of Brandom's proof, to summarize it briefly, is that deontic scorekeepers during their exchange of reasons in a pragmatic language game ought to reflect on the different perspectives of other scorekeepers and adopt a perspective-transcendent view that focuses on the fundamental 'aboutness' of their exchange. If they did not translate the utterances of others into a propositional language that utilizes an implicit notion of the mind-external world, their discourse would not be possible and fail. They would only talk about 'what' others have claimed, and thus their communication would be meaningless. Here, the idea of communication means to take over the perspectives of others by substituting elements of their utterance and thereby connecting the content of their utterances to one's own semantic net. For explicating this idea, Brandom utilizes the whole inferentialist repertoire that he has developed before. In particular, he relies on the inferential, substitutional, and anaphoric articulation of contents on the semantic side, and the normative terms of commitment and entitlement on the pragmatic side.

As already discussed, Brandom's model is based on different pragmatic components that are necessary to establish an intersubjective, attitude-transcending meaning during our exchange of reasons. These components are so far:

a the intersubjective, responsible play of deontic score-keepers with assertions

b the inferentialist-normative structure of their moves during their play

We cannot simply interpret these formal (ahistorical) requirements, as part of a mind-external realism yet, since each of them is understood as an internal, "structural aspect" of our currently played language games (Brandom 1998, 597). However, Brandom identifies his semantic model as a "semantic externalism" according to which meaning "depends on how things actually are, whether we know how they are or not" (Brandom 1998, 647; see Whiting

---

transcends the perspective of the involved deontic scorekeepers. Prien and Reinikainen make some suggestions on how to improve the arguments of *Making It Explicit*. In this regard, I will look at different, later forms of Brandom's proof after I reconstructed his approach in *Making It Explicit*.

95 Brandom, Robert. "Replies." *Philosophy and Phenomenological Research* 57 (1), 1997: 189–204. doi:10.2307/2953789.

2008, 584).[96] Yet, at the same time, Brandom's 'semantic externalism' is inherently perspectival. Facts are contingent upon the current perspective of the scorekeepers and cannot be objectively assessed from an external viewpoint (see Brandom 1998, 647). Nevertheless, this is not a problem for Brandom, who claims that "semantic (perspectival) externalism begins at home" (Brandom 1998, 647).[97] He aims to provide an account of how mind-external reality constrains our discourses from within the pragmatic language game. As Brandom claims, the semantic "external interpretation" is part of an "internal interpretation" of scorekeepers (Brandom 1998, 647). Now, he himself raises the critical question: "what makes one interpretation better than another" (Brandom 1998, 647)? Brandom's answer is that it depends on the "practical norms of understanding one another here at home" (Brandom 1998, 647). In other words, the norm for good discourses originates with the speakers themselves, who seek to engage in productive discourses. To fully understand Brandom's internal approach, we must grasp the concept of the discourse-participants as scorekeepers on a deeper level.

The introduction of a shared perspective is directly tied to the challenge of communication. As previously discussed, the holistic communication framework encounters an issue wherein each participant, or scorekeeper, possesses their own conceptual content based on their individual inferential network. Surprisingly, Brandom views this divergence in individual semantic networks as the foundation of our communication process. If we all perceived and communicated about the world in exactly the same manner, there would be no need for communication. The divergence in perspectives serves as the essence of communication, as Levine aptly highlights: "[...] without other scorekeepers attributing normative statuses to them, whatever commitments and entitlements that seem right to them would be right" (Levine 2007, 9). The idea of Brandom's reciprocal deontic score-keeping, therefore, introduces the other as an instance of necessary correction and so avoids the first person error, i.e., to

96 Whiting questions whether Brandom is really a semantic externalist: "In fact, on closer inspection, Brandom's version of IRS [inferential role semantics] does not in any way appear to provide genuine externalism. According to it, the meaning of an expression or the content it expresses is determined by how things objectively are according to the one ascribing or attributing that meaning or content. It is, then, still the case that, in Brandom's theory, meaning is determined in the first instance by the attitudes of each speaker, not by extra-linguistic reality" (2008, 593). Whiting does not think that Brandom's inferentialism can give an account of mind-external reality (see Whiting, Daniel. "Meaning Holism and De Re Ascription." *Canadian Journal of Philosophy* 38, 2008: 575–599. http://www.tandfonline.com/doi/abs/10.1353/cjp.0.0033).

97 Probably alluding to Quine's famous claim that 'translation starts at home'.

assume that something is right just because 'I' claim something to be right. As Brandom puts it in his more formal language, the involvement of others, who may have other views, is necessary since only then "it is possible that (*I* believe that *p* and it is not true that *p*)" (Brandom 1998, 604).

The dialogical account of meaning production, however, does not have to be related to an external, mind-independent reality, yet. Since we introduce the game of truth obliging each other to play this particular game, truth so far depends on each particular game and the limited historical, conceptual net of the involved players. Brandom now needs to overcome the problem that a community establishes correctness according to their 'community standards'. In his words, he wants to answer how "objectivity precipitates out of the social soup of norms that are whatever the community takes them to be" (Brandom 1998, 54). In order to achieve this, he claims that we can distinguish between the "applications of concepts that are objectively correct and those that are merely taken to be correct" (Brandom 1998, 595). With regard to this, he introduces the distinction between de dicto ascriptions and de re ascriptions that we as language-users keep track of in different books. The de dicto book includes commitments the scorekeeper asserts (see Reinikainen 2021, 60–61, Chung I-Lin 2008, 107).[98] The de re book collects the consequents and antecedents that the speakers must commit to on the basis of inferences (see Chung I-Lin 2008, 107). 'De dicto' stands for the statement *of* somebody, while 'de re' stands for what the statements are *about*. According to this double-entry accounting, contents do not only depend on people but also depend on the objective consequences that they entail. Brandom's idea is now that scorekeepers mediate their different views by substituting subsentential parts in their assertions.[99]

98 Chung, I-Lin. "No Communal and First-Person Errors: A Critique of Brandom's Objectivity Proof." *National Cheng Chi University Philosophical Journal* 20, 2008: 105–122. Accessed on November 26, 2017. https://thinker.nccu.edu.tw/upload/40/publication_file/551/林從一(Chung-I%20Lin)20.pdf.

99 This distinction bears resemblance to Hegel's examination of the in-itself and for-itself. However, Hegel emphasizes that this distinction is created by consciousness itself, leading him to conduct a dialectical investigation. This approach culminates in a circular, epistemological strategy that expands into an anthropological exploration of self-consciousness and its historical development within society as spirit. Hegel's strategy is fundamentally circular as it requires us to comprehend the differentiation between an assumed, external reality and reality as experienced by consciousness alone. Hegel does not advocate for knowledge of a reality independent of the mind but focuses on truth in relation to human understanding. In contrast, Brandom introduces this distinction without engaging in such reflection. His framework is foundational, asserting that these structures are necessary for meaningful discourse. However, beyond Brandom's foundationalism, the exact meaning of 'objectivity' remains unclear. Hegel addresses this problematic term in his chapter on objectivity in the *Greater Logic*, agreeing on its necessity.

The substitution is governed by the *salva veritate* rule. It means that the substitution of terms within an assertion must preserve the meaning of the assertion. Moreover, the entries in our de re book should be without incompatibilities. Brandom uses the following example for explanation:

> DE DICTO: he believes that drinking the liquor distilled from the bark of that kind of tree will prevent malaria.
>
> DE RE: he believes of quinine that malaria can be prevented by drinking it.
>
> see BRANDOM 1998, 513–514

So in the de dicto-example, we attribute the commitment that liquor from the tree prevents malaria. In the de re-translation the phrase 'liquor distilled from the bark of that kind of tree' is replaced by 'quinine'. By using the concept 'quinine', we invoke other inferential relations that confirm that the concept works within our semantic net without contradiction. In other words, the concept is compatible. This describes our capacity for rehearsing and taking over others' perspectives: "if you want to understand what I say, you have to be able to associate with it a sentence that in your mouth expresses the same claim as the sentence uttered expresses in mine" (Brandom 1998, 510). By using the de re description, we are transforming an otherwise unknown object and exploring its meaning inferentially in our "mouth" (see Brandom 1998, 514). So when we have another de dicto utterance like 'he says that the seventh god graces us with his presence' (example from Brandom 1998, 514), the utterance may be unclear to us because we cannot make sense of the term 'seventh god'. As a result, it has "consequences" that are impossible to "endorse" because the utterance produces no object for us (see Brandom 1998, 514). Nevertheless, inferential exploration of different consequences can bring us to a more specific 'de re' formulation: 'he claims of the sun that it is shining' (example from Brandom 1998, 514). This sentence that substitutes an unknown term without significance receives its significance by virtue of a transformation. We can check it with the overall compatibility with other contents in our semantic net and also check it with regard to other de dicto statements and attached consequences. As Brandom says: "Identifying what is being talked *about* [...] permits

---

The argument aligns with Brandom's argument for objectivity, as subjectivity confined to itself would render its meaning futile, making it impossible for a subject to assert anything. However, Brandom presupposes that objectivity concerns the connection between our statements and a mind-independent reality. Moreover, he contends that we must communicate through propositions to convey meaningful content.

me to extract information across a doxastic gap" (Brandom 1998, 514). In other words, we can generate knowledge on the basis of different perspectives and expand thereby our semantic nets inferentially.

An accepted statement in the books increases the knowledge we have of a presupposed, mind-external world. In order to now explicate the role of the mind-external world in this construct, Brandom translates the classical conditions for knowledge into his inferentialism:

a Belief is here the attribution of an inferentially articulated, propositional commitment (see Brandom 1998, 515).
b Justification is the attribution of entitlement (see Brandom 1998, 515).
c Truth is the normative action of the scorekeeper undertaking "the same propositional commitment attributed to the candidate knower" (Brandom 1998, 515).

Now, as readers who wrestled with Brandom's terminology and followed his vast preparation for this ultimate argument, we probably had expected a great revelation for the aspect of truth. It may be a bit disappointing that truth consists only in the norm of us committing to the same proposition that we attribute to others. But Brandom explains that truth in general is not about "attributing a property" to an object (Brandom 1998, 515). Rather, truth in our discourses means to take over a "*normative stance* with respect to a claim" (Brandom 1998, 15). If I understand him correctly, this normative stance means that I accept that the others' perspective of something is, at the same time, a perspective for me. To express it in more metaphysical words, we ought to transcend ourselves. And to say it in less metaphysical words, we ought to generate a position that can include different perspectives consistently. Ultimately, it is, however, not about the particular, generated, new perspective, but about the normative stance that we direct ourselves in the same direction. In this discursive direction lies something beyond our current perspectives. It sounds pretty much like a linguistic, intersubjective version of Kant's transcendental x. It is a common point of convergence that constitutes a common surface according to which new knowledge in terms of inferential connections can be generated. Objectivity consists then in acknowledging the norms that are necessary for this kind of communication that converges in a presupposed transcendental x. So the "key to objectivity" lies in understanding the norms as conditions under which a statement can be true for me and for the other (Kusch 2002, 254).[100] In Brandom's words, "truth talk properly" understood means then to know and

100 Kusch, Martin. *Knowledge by Agreement.* Oxford: Clarendon Press,, 2002.

bridge the perspectival difference "between *attributing* a normative status [... and] adopting" the status (Brandom 1998, 515).

The question remains how this norm of generating common perspectives justifies a relationship to 'real things'. I must guess here. As there are infinite perspectives possible, the scorekeeper's book is potentially infinite. Yet, the more perspectives the scorekeeper includes or excludes from this book, the more reliable the content consisting of propositions, and the more reliable becomes the book. Truth, as the normative stance of 'looking into the same direction', is about describing the same 'thing' consistently from different angles. It is a perspectivist idea that the more we describe it, the better.

Is this a metaphysical, mind-external realism? Not explicitly, but I think we can assume here that mind-external reality must be somewhat stable and cannot change for Brandom's formal, perspectival system to work. Only if there is a stable, mind-external reality, we have a reliable base for the perspectival transformation of the score-keeping method. Before I discuss, however, my main objection, I will introduce some common criticisms against Brandom's approach.

Objections regarding Brandom's Proof of Making It Explicit: Kusch evaluates Brandom's proof: "Contrary to what Brandom alleges, he has not shown that objects are 'non-perspectival in a strong sense'" (Kusch 2002, 258). According to Kusch, the non-perspectival perspective allows only for a new perspective. As Kusch puts it: "The absence of perspective is thus itself only perspectival" (Kusch 2002, 258). However, Brandom would not deny that it is a new perspective. So Kusch's criticism misses a bit of Brandom's aim. Brandom believes that we approximate reality by continuing this method of destroying false perspectives and thereby tightening our semantic net. In that regard, Brandom is fully aware that perspectivalness will always remain. To say it again, the norm is that we are always willing to correct our perspective on the basis of the ideal, propositional transformation of perspectives.

Huetter-Almerigi objects: "Yet, what ultimately accounts for objectivity is the difference between our perspectives and the non-perspectivalness of the world beyond us. Our claims are correct if they get reality right (reality without us), although we are in Brandom's 'sanitized version,' as he calls it, never directly in touch with it" (2020, 8). I think that also Huetter-Almerigi's criticism misses the target slightly. Brandom's approach is not explicitly about getting things right, but about establishing a norm that guarantees progress towards mind-external reality. In other words, we get something about facts right, but it is unclear what this is in its entirety. Objectivity consists then of the fact that we eventually will weed out more and more false beliefs.

Reinikainen, like many other interpreters, objects that Brandom's proof "leaves entirely open the crucial question of the source of semantic correctness" (Reinikainen 2021, 63). As it should be clear by now, however, this is intended and due to Brandom's *pragmatism*. We can never exhaustively define the source of our concepts. It, therefore, is not Brandom's intent to clarify what mind-external reality is. For a more precise criticism, however, Reinikainen develops a helpful distinction. Regarding the question of what Brandom intends to show with his concept of objectivity, he distinguishes the following theses:

1 First, Brandom may mean that "a norm n is attitude-transcendent for community C iff it is possible for everyone in C to be mistaken about the correct applications of n" (Reinikainen 2021, 63)
2 or Brandom may mean that a "norm n is properly objective for community C iff the world of objects partially determines the correct applications of n" (2021, 64).

I would like to add a further stronger criterion:

3 Brandom may mean that a 'norm n is properly objective for community C iff the world of objects *reliably* determines the correct application of n'.

I think that Brandom intends to prove the latter (3), but has arguably proven the first (1), could, however, also defend the second (2). Reinikainen believes that Brandom defends (1), but has not yet defended (2). He believes that Brandom, however, can defend (2). It requires the concept of conceptual realism that Brandom, however, explicated only after *Making It Explicit*. Now, on the basis of his distinction, Reinikainen concludes that "everyone in a community may intelligibly [...] be mistaken about the correct application of a norm", which, however, does not imply "that it is the world of objects which determines [...] the correct applications" (Reinikainen 2021, 64). As I evaluate it, we have only distanced ourselves explicitly from the community view (1), which probably involves the fact that reality is somehow involved in the correction of our beliefs (2). However, do we really doubt that reality is somehow involved in our communication? I do not think we doubt this. Of course, we all somehow respond to reality. For example, in case we see white mice everywhere, we still respond to the physical fact of a malfunctioning brain. So the question remains whether we can know that reality allows us to change our views reliably so that a conceptual approximation toward mind-external reality is possible. My general objection, therefore, is that our practices are not necessarily arranged with regard to how things really are, but according to how we take things to be in an experiential, historical process. The 'real' thing is, however, normatively charged. It means that the conceptual shape of what we take to be reality is only a theoretical conception within our specific normative framework. Thus, access to it is generated through a combination of the limited experiences of humans. As I see it, every perspective is so far relative to this framework based on experiences. Of course, Brandom seeks to avoid this relativism by insisting that: "Discursive practices incorporate actual things. They are solid—as one might say, *corporeal.* They involve real bodies [...] The way the world is constrains proprieties of inferential, doxastic, and practical commitment in a straightforward way from *within* those practices" (Brandom 1998, 332). But again, I think, there is really no reasonable doubt that reality is somehow involved in our discourses. In this sense, for example, nobody in the current debate seriously denies that reality exists. Yet, the difficult question regarding reality is how we can guarantee that we get closer to it. I do not think that Brandom shows this with his proof (probably this is also the criticism Reinikainen has in mind). Rather, Brandom shows that we can correct our views, once our experiences do not align with the formal semantic net we have developed in our communities. Then, we can change it within our community and recognize that we have erred. For this, we, indeed, need a method that we can agree on to be objective. I believe that Brandom gives us a good model for this. Nevertheless, the new views we generate on the basis of this method do not have to be closer to reality. We only exclude false perspectives from many possible perspectives. Our new perspective, however, may still be false and may misguide us even more, and strangely enough, our former rejected belief may turn out to be true in a modified sense later again.

Let us consider another objection. I agree with Brandom that knowledge does not lie in grasping things as mere content, but rather in focusing on the correct process we should follow. However, doesn't the process we derive also involve expressing it through a semantic concept? Nevertheless, according to this approach, semantic concepts never fully capture reality. This brings us back to the inferentialist circle, which seems to perpetually loop back to its origin. Therefore, even if Brandom's proof successfully demonstrates that reality consistently corrects our perspectives within the framework of inferentialism, the problem arises when we attempt to extract the method, as it results in a limited and potentially erroneous semantic concept. One could argue that we still get closer to reality, even with an incorrect method, because we eventually correct our semantic concept. However, the issue here is that if the method is wrong, we will not reach the correct method through the wrong one—any progress would be purely coincidental. There is no reliable pathway from an incorrect method to the correct method through the incorrect method itself.

So, I think the criticism has also to be directed toward the semantic phrasing of the procedural conditions. When Kusch, therefore, criticizes that Brandom's communally achieved "absence of perspective" remains "itself only perspectival", this criticism has to be rather applied to Brandom's semantic extraction of a procedural method (Kusch 2002, 258). The semantic concept of the method is only perspectival. It cannot escape relativism.

However, I would like to interpret the relativist result positively. With respect to the method, I would not say that the method is 'only' perspectival. As I would interpret it a bit more positively: we have, indeed, achieved a historically higher perspective. In this sense, Brandom transforms, first of all, the historical problems of an analytic stream of philosophy into inferentialist, normative pragmatics. He then proceeds to turn inferentialism's perspectivism against itself and sublates it into a higher form of perspectivism. It can justify itself to be a historically better truth with regard to its predecessors following a community-internal standard, and further develop internally. Moreover, instead of adhering to perspectival skepticism, it allows for an objective process of the correction of concepts within its community that ultimately can transform the community into a new community.

So if we now take Brandom's results as a general stance on truth, we can see it is a relativist perspectivism. We never arrive and our generated truths depend on our 'community standards'. The advantage, however, is that by virtue of its open stance, we remain open for our views to be corrected so that one day, we may, indeed, reach the truth. I think this is what relativism can achieve: an open stance towards truth.

Nevertheless, for Brandom, to say it a bit bluntly, the gained perspective consists of a designed, sophisticated, formal, and finished discourse machinery that ultimately presumes that we get closer to the mind-independent reality and which will lead us—also through the historical works of Hegel—to a better future. In this future, we will be, as Brandom emphatically announces, "magnanimous" people (Brandom 2019, 32). As I read Hegel in that regard, however, not only content but also such machinery is the result of a preliminary plateau that cultures can achieve, but only within history. This plateau does not let us approximate mind-external reality. The plateau may even turn out to be false, or, it may be eventually sublated. But with this achieved position, we can, on the one hand, manipulate 'facts' of our experience; on the other hand, we maintain an open stance for a possible ontological truth that may arrive one day. Since we, in so far as we are relativists, do not attach ourselves to a false theory, we then may be capable of acknowledging it.

Given my interpretation, it is interesting to read Brandom's approach through a more relativist lens. Rorty, for example, recommends in a similar way to 'switch' from "progress toward a focus imaginarius to improvement on the historical past" (Rorty 1997b, 175). So instead of delivering objective standards with regard to a mind-external reality, we work out the rules of how we want to understand ourselves with regard to what we commonly agree to experience and according to what we perceive as a viable path for humans. As I already said before, the restricted perspective on human reason may then be replaced with a stance on the openness of humans to develop. I believe this is Hegel's idea of human freedom. It means that we are not tied to a theoretical conception of ourselves, but can grasp ourselves as free and so become ourselves. Now, the historical process of abstraction and formalization is thereby not to be dismissed, but we should be aware that these formalizations can be overturned by new forms and contents. We therefore may use Brandom's discourse machine for a while, but we certainly treat it as a machine that can depreciate in value. I, therefore, agree with Rorty here: "This amounts to switching from pride in being closer to Reality to pride in being farther from the cavemen" (Rorty 1997b, 175). As he sees it, it is a "retrospective celebration of the increasing Muendigkeit of our discursive practices, of the increasing self-consciousness and sense of responsibility which has emerged from the last five hundred years or so of Western history" (Rorty 1997b, 175). If I understand Rorty correctly, Brandom actually directs us towards such a more relativist philosophy that then, according to Rorty, would even "follow in the footsteps of Hegel" (Rorty 1997b, 177). This is a philosophy of radical freedom, which means that we are ultimately not bound to formal assertions even though they are tools free to use.

All relativist interpretation aside, Brandom turns more towards a mind-external realism instead of taking Rorty's recommended path. So, it is a different Hegel that Brandom will reconstruct for us. Before we come to his Hegel, however, there still remains the question of what is actually Brandom's mind-external realism. I think this mind-external realism becomes more pronounced in the proofs he develops after *Making It Explicit.* I will therefore look at one of his later arguments and connect it back to *Making It Explicit.*

A Further Proof: Brandom's proofs for objectivity in relation to a mind-independent reality are further explored and modified in *Articulating Reasons* (2000).[101] In *Articulating Reasons* Brandom writes how things are "transcending the attitudes of practitioners" and further defines the major task:

> So the challenge for assertibility theories is to start with a notion of propriety of assertion that is grounded in and intelligible in terms of the practice of speakers and audiences, and yet which is rich enough to fund

101 Brandom, Robert. *Articulating Reasons: An Introduction to Inferentialism.* Cambridge: Harvard University Press, 2000a.

> normative assessments that are objective in the sense of transcending the attitudes of practitioners.
>
> BRANDOM 2000a, 198

For Brandom, assertibility theories face the problem of an incompatibility-difference, if equal propositional sentences are transformed into an explicit formulation. To explain 'incompatibility difference' Brandom uses the following examples:

1. The swatch is red.
2. The claim that the swatch is red is properly assertible by me now (2000a, 198).

The two claims are related since the second claim is an explicit version of the former one. Thus, as Brandom argues further, the assertibility conditions are the same. The truth conditions, however, are not the same (see Brandom 2000a, 199). Brandom claims that if both statements were the same, then both of them would be equally incompatible with other claims. In the case of the claim 'rational beings never evolved', however, the first claim that the swatch is red would be compatible with it, while the second claim could not be stated (Brandom 2000a, 199). Obviously, humans have to exist in order for this claim to exist. The difference between proposition (1) and its explicit version in (2) poses a problem for assertibility theories. It challenges the feasibility of developing a theory of assertions that can consistently explain how to make a proposition and its context explicit. Regarding this problem, he states:

> the commitments and entitlements they associate with ordinary empirical claims such as 'The swatch is red' generate incompatibilities for these claims that differ suitably from those associated with any claims about who is committed to, entitled to, or in position to assert anything.
>
> BRANDOM 2000a, 204

In order to solve this problem, Brandom wants to show that objectivity does not depend on the community that makes it true or false. It involves further conditions that constitute the possibility of objective errors of the speakers. For Brandom, these further conditions depend on how things really are. He writes on this objective entailment: "those contents display *objectivity* of a particular sort: they are not about any constellation of attitudes on the part of the linguistic practitioners who produce and consume them as reasons" (Brandom 2000a, 190). Brandom suggests that there are incompatibilities that are independent of utterances, while he wants to discuss these context-invariant truths again through his distinction of de dicto and de re. To repeat, de dicto are claims that I am committed to and de re claims are claims that I am entitled to by virtue of objective entailments of the included propositions. Brandom claims that the original sentence 'the swatch is red' cannot be simply transformed into the assertion of claiming without noticing that it changes from a de re claim into a de dicto claim. This is because our original claims are, according to Brandom, about mind-independent reality. Hence, they should be discussed as de re claims and not as de dicto claims. In de re claims, mind-external reality appears first as propositional content:

> The point of all this is that the *objectivity* of propositional content [...] is a feature we can make intelligible as a structure of the commitments and entitlements that articulate the use of sentences: of the norms, in a broad sense, that govern the practice of asserting, the game of giving and asking for reasons.
>
> BRANDOM 2000a, 203

So Brandom takes the incompatibility-inadequacy as an indication of the fact that we refer in our statement to objective propositions. These propositions that we then utilize in de re claims are distinct from our de dicto claims. It makes use of the concept of objectivity. By 'objectivity' he understands then "something that could be true even if there had never been rational beings" (Brandom 2000a, 203). In Brandom's words, this means more explicitly "that in claiming that the swatch is red we are not saying anything about who could appropriately assert anything, or about who is committed or entitled to what" (Brandom 2000a, 203). For Brandom, the content of a claim is different from our navigational efforts and the position of the speaker. According to his argument, the objectivity of our claims comes from using propositional content. So, what really makes something objective is when we collectively recognize a reality beyond our minds, expressed through propositions.

However, the argument remains unfinished as it presupposes our discussions involve mind-independent objects. It does not demonstrate the existence of the specific mind-independent objects referenced in our statements. Brandom, therefore, argues further that there must be something that grounds our assertions. This must be the way of how things really are. Brandom uses for this proof his normative vocabulary of commitment and entitlement. He first observes that the simple 'enhancement' of the explicit form of our sentences does not alter incompatibility-differences:

> 2'. I am now committed to the claim that the swatch is red.
> 2". I am now entitled to the claim that the swatch is red.
>
> BRANDOM 2000a, 200

Both of these claims are still incompatible with the claim that 'rational beings never evolved'. Nevertheless, they indicate how the speaker's position in a community depends on more conditions than the conditions for asserting. The first claim depends on the speaker's attitude, while the second claim depends on the reasons why the speaker believes to be entitled. With regard to this, Brandom claims:

> Put another way, looking at propositional content in terms of incompatibilities, themselves defined in terms of the fundamental normative statuses of commitment and entitlement, provides the expressive resources to distinguish between the sense of 'assertible' that falls short of guaranteeing truth [...] and the sense [...] that would guarantee truth.
>
> BRANDOM 2000a, 203

So, secondly, by using the normative vocabulary of commitments and entitlements we can, according to Brandom, identify something beyond assertions that guarantees truth. For our statements to work the way they do, we have to *presuppose* something that is independent of rational beings. For example, there is a "*consequential commitment*" so that the interpreter of the claim can conclude from the assertion 'the swatch is red' that 'the swatch is colored' (Brandom 2000a, 191). He explains this with regard to parrots. Parrots respond by virtue of a stimulus. There is no further consequential commitment to the fact that the swatch is colored. Sapient animals, on the contrary, answer in terms of inferences. We, therefore, agree that something follows from our assertions. Yet, this is still not proof of a mind-external reality. We only presuppose a commonly shared world, a world that is produced mind-internally and object within our communication. In other words, we agree on something mind-external mind-internally. However, as Brandom argues further, it follows that assertions possess "inferential *consequences*," and "inferential *antecedents*, relations to contents that can serve as premises from which entitlement to the original content can be inherited" (Brandom 2000a, 193–194). So, we are committed to an assertion because it entails something and we are entitled to our assertions because they depend on antecedents. The idea must be that without such commitments to entailments and without entitlement by virtue of antecedents, we would not have assertions that could be taken seriously. We would not be able to play language games. As I understand Brandom, our assertions somehow inherit a particle of mind-independent reality.

Thirdly, we have to consider that in order for something to count as an antecedent or an entailment, it must be propositionally structured since otherwise, we could not pick it up in our assertion. Brandom develops here his concept of conceptual realism.[102] For our discourse to work the warranting factors of our statements ultimately relate to a mind-independent reality that must be structured conceptually, otherwise we would not be able to discuss anything at all. In other words, the mind-independent world is the ultimate beginning of all communication from which we inherit the propositional structure of our statements. So Brandom deduces a concrete structure of mind-independent objects since otherwise our objective discourse of propositional assertions would be impossible.

Allow me to propose a reconstructed version of this transcendental argument as follows: Firstly, we must establish objectivity within our exchange of reasons. During communication, the speakers have to ask: what does the other speaker infer in relation to the content of their statements? Secondly, for effective communication, the exchange of reasons necessitates the inclusion of propositions. Without propositions, our statements would be incomprehensible to one another. Lastly, in order to engage in any meaningful discussion, there must exist an underlying reality as a starting point that possesses propositional structure, which all our propositional statements inherit. From the fact that we objectively communicate through propositions, it follows that mind-independent reality must be structured propositionally.

To reiterate: Brandom asks for the circumstances that entitle us to claims (see Brandom 2000a, 196). These conditions are not context-variant but are assumed as stable, based on mind-independent facts since without these there would have been no rational way to communicate. In other words, all rational communities must follow the pattern of assertive speech. According to Brandom, our assertions are thus attitude-transcendent. Moreover, objectivity hinges on the premise that our engagements with commitment and entitlement rely on a propositional syntax mirroring the world's structure. Without such syntax, we could not be rational. This syntax would, moreover, lack rationality if it failed to correspond with reality.

---

102 As Reinikainen points out, conceptual realism is not yet an explicit term in *Making It Explicit*: "The missing piece for Brandom's pragmatist quest for semantic objectivity is conceptual realism or the idea that both subjects and objects can be understood as conceptually structured. While conceptual realism only comes into explicit focus in Brandom's later works" (Reinikainen 2021, 56). As he elaborates, the concept of conceptual realism would already solve some of the issues regarding Brandom's proofs of objectivity in *Making It Explicit*.

At this point, it remains unclear why propositions must solely derive their meaningfulness from a mind-independent reality. Brandom asserts that objectivity is unattainable unless our discourses are connected to a propositionally structured, mind-independent reality. But perhaps the specific objectivity he insists upon is simply unattainable. I will delve deeper into this matter later, especially in the context of Brandom's interpretation of Hegel (part 2.4).

In *Making It Explicit*, we find further supportive arguments. So, for example, he criticizes a mere phenomenalism that cannot successfully establish a relation to mind-external reality (see Brandom 1998, 294). He then expresses this with regard to the world being a constellation of facts, while our factual claims "are structured and interconnected by the objects they are facts about; they are articulated by the properties and relations the obtaining of which is what we state when we state a fact" (Brandom 1998, 333). Brandom believes that our commitments only make sense if they are about objects that are "propertied and related" (Brandom 1998, 333). In other words, even though all of our assumptions about the real object might be wrong, the 'fine structure of normativity', i.e., endorsement, commitment, and entitlement, indicates that there is a mind-external reality that is propositionally structured (see also Bransen 2002, 382).

Bransen extends Brandom's claims even further:

> if one is entitled to assert 'Thunder will be heard soon' or 'The streets will be wet' or 'That is a door', then this means, truly and objectively, that there is thunder and that it will be heard soon, that there are streets, and that they will be wet, and that there is a door.
>
> BRANSEN 2002, 383

I do not disagree that we have phenomenal experiences. For instance, if an alcoholic claims to see white mice, it is likely that they genuinely perceive white mice. Their subjective experience warrants their claims. However, this does not imply that white mice objectively exist in mind-external reality.

Instead, I propose that we lean towards a form of discourse characterized by 'seems-talk'. When I use phrases like 'There seems to be a door', I deliberately avoid making definitive assertions. By incorporating the qualifier 'seems', I acknowledge the inherent uncertainty in our experience. This approach allows me to refrain from dogmatic assertions and instead invites a more open consideration of reality, which is a form of relativism. We can then develop theories based on shared experiences. These theories remain susceptible to change, as our experiences themselves can undergo fundamental shifts. On the contrary, Brandom proposes a mind-external realist stance as an escape from global alethic relativism. According to him, if we constantly withhold judgment and apply 'it seems' to all endorsements, we arrive at the absurd consequence of 'something seems to seem'. At least, according to Brandom, we had to accept that there is something that seems to be. Against positions like mine, Brandom, moreover, argues that withholding of an endorsement (it only seems) stems from a former experience of failed beliefs (see Brandom 1998, 293). Seems-talk, therefore, presupposes an original aboutness. Brandom thinks that this is where the "subjective phenomenalism" of classical pragmatisms fails (see Brandom 1998, 212). In particular, he elaborates that 'seems' is a result of experiential episodes:

> these locutions are introduced after such a practice is under way, as a way of dealing with systematic sources of perceptual error that one becomes aware of through having to withdraw unreflective, noninferential claims on the basis of their incompatibility with commitments one is otherwise entitled to.
>
> BRANDOM 1998, 293

In other words, saying that something seems a certain way comes from an overall experience that things can be different from how we take them to be. 'Seems' expresses a "withholding" of an endorsement (see Brandom 1998, 293). 'Seems' presupposes a common understanding of something being genuine, as he elaborates: "Because in order of understanding, grasp of what is expressed by concepts of the form 'seems-K' presupposes a grasp of what is expressed by corresponding concepts of the form 'is-K'" (Brandom 1998, 295). This realization of a connection between semblance and reality is not a novel revelation; it bears, indeed, resemblance to Hegel's observation that something exists both in the realm of consciousness and can also be considered in itself. However, Hegel delves deeper to unveil the dialectical relationship between these perspectives, contending, moreover, that the distinction itself is a construct within consciousness. In contrast, Brandom's assertion that seems-talk is based on a concept of mind-external reality overlooks this intricate dialectical interplay. Brandom simply presupposes the primacy of the notion of mind-external reality, a perspective that will later influence his interpretation of Hegel's *Phenomenology*.

Against Brandom's perspective, one could argue that even the term 'reality' loses its significance without the possibility of semblance. In essence, the concept of reality would not be discussed if semblance were not always a possibility. From this standpoint, both terms are intricately interlinked in a dialectical manner; neither precedes the other. Consequently, they require a mutual grounding to be adequately understood and discussed. Therefore, phenomenalism, which Brandom explicitly dismisses, can be considered as an, at least, equally valid option compared to Brandom's conceptual realism. Rather than outright rejecting one in favor of the other, I hold the view that it is necessary to explore the dialectical relationship between phenomenalism and mind-external realism concerning the experiential aspects of consciousness.

Furthermore, Brandom claims that for each phenomenalist claim, there is a corresponding "*realist*" (Brandom 1998, 296) claim since semblance is grounded in 'real' experiences. Brandom seeks to replace phenomenalism with an explanation centered on linguistically "undertaking and attributing commitments" (Brandom 1998, 297). Brandom thereby simplifies the issue of authentic phenomenology into a linguistic ambiguity that can be resolved through the "fine-grained" delineation of commitments and entitlement (Brandom 1998, 297). Against this, I argue: while 'seems-talk' typically arises after experiencing disappointments and a pursuit of genuine truth, 'truth talk' emerges as well after recognizing the possibility of disappointments.

In my view, Brandom's proofs do not convincingly demonstrate a reliable connection to mind-external reality; rather, they illustrate how we justify and correlate our endorsements with a particular conceptual framework that Brandom articulates in terms of commitments and entitlements. This framework is a formal, linguistic model, which, while valuable, does not provide a definitive solution to the challenge of reliably approximating mind-external reality. However, I want to emphasize again that I do not reject the term of reality. I argue, however, that reality is a subordinate concept that needs to be reconstructed within a historical model of how we express ourselves during the course of our historical development with respect to our experiences as humans. This aspect will become more prominent once we discuss Hegel's *Phenomenology*.

Before, I pointed out how Rorty was first in favor of Brandom's inferentialism that avoids notions of a classical 'unconditional'. He thought that Brandom would be in agreement with Davidson's rejection of truth-makers. Yet, now, realizing how Brandom uses the terms 'getting right,' 'really is' and 'making true', he thinks that Brandom's ideas "would make Davidson's hair stand on end" (Rorty 2021, 170). He particularly believes that these ideas will become tools in "authoritarian hands and be used for reactionary purposes" (Rorty 2021, 173). Brandom's model, according to Rorty, opens the door for abuse, in particular, for the idea that reality has normative influence on our discourses. He believes that Brandom "countenancing these dangerous idioms will be taken as a concession by the bad guys" (Rorty 2000, 187). 'Bad guys' are "authoritarians" who, for example, understand "perceptual experience as a model for 'hard facts'" (Rorty 2000, 187). He warns Brandom that these 'bad guys' do not operate in the same paradigm as him according to which "increased freedom and richness of the Conversation" is our shared goal. Like Brandom, they rather "aim [... at] getting Reality right" (Rorty 2000, 187). However, Rorty, probably also speaking in friendship, acknowledges that Brandom's "heart is certainly in the right place" (Rorty 1998, 135). So, there is no doubt about Brandom's genuine intentions. My disagreement with his philosophy is therefore grounded on his emphasis on a mind-external reality and the formal discourse model. It is incompatible with Hegel's, as I see it, better, historical relativism. Here, I share Rorty's push for a conversational, humanist discourse, according to which we improve our understanding of which world we want to inhabit.

3 Brandom and History

I have not yet discussed Brandom's view on Hegel since Brandom's *Making It Explicit* does not mention him. He introduces his interpretation of Hegel in *Tales of the Mighty Dead* (2002)[103] and in *Animating Ideas* (2009). There, he argues that Hegel is an inferentialist with holistic intentions. Fitting Hegel to his own theory, he mostly utilizes the chapters of Hegel's *Phenomenology* that he can identify as semanticist. While Brandom's application of his semantics to Hegel is questionable, my primary criticism lies in his lack of awareness regarding the broader issue of history. One of the few passages discussing the concept of history begins with a, in my mind, problematic definition of noumena and phenomena that he later also applies in *A Spirit of Trust* (2019):

> [... Hegel's] answer is that the idea of <u>*noumena*</u>, of things as they are in themselves, the reality that appears in the form of *phenomena*, can be understood practically in terms of a distinctive role in a recollectively rationally reconstructed historical sequence of phenomena.
>
> BRANDOM 2009, 100

Brandom grasps history as a secondary phenomenon that serves us to understand the relationship between phenomena and a noumenon, for Brandom, to the way how things really are. This interpretation is based on Brandom's view of the relationship between phenomena and noumena:

---

103 Brandom, Robert. *Tales of the Mighty Dead: Historical Essays in the Metaphysics of Intentionality.* Cambridge: Harvard University Press, 2002.

> One of the senses in which what he presents is a *phenomenology* is that he starts with an account of phenomena (what things are for consciousness) and seeks to reconstruct the notion of noumena (what things are in themselves) out of the resources it provides.
>
> BRANDOM 2009, 100

I regard this as questionable. As I understand Hegel, he does not attempt to recognize the noumenon or to give a concept of it. Instead, Hegel attempts to answer how a phenomenology without a noumenon can be objective. According to Brandom, however, the *Phenomenology* is about the "rational integration into the constellation of one's prior commitments of some new commitment [... which ultimately] is intelligible as one's commitments as to how things really are, objectively, in themselves" (Brandom 2009, 100). So our main intention is to relate our discourse activity of uttering commitments to how things really are. History is then a progression of commitments that are replaced by new commitments ultimately approximating mind-external reality. In other words, there are no paradigm shifts in one rational history, but there is one evolving net of semantic commitments towards mind-external reality. The underlying basis for this is the ideal form of communication—the exchange of reasons—that essentially never changes during the course of history (see Brandom 2009, 101). Brandom describes this relationship to past commitments as 'a kind of rational recollection': "This means—as being what one takes to be not just an appearance of that reality, but a *veridical* appearance, one in which things appear as they really are—when it is accompanied by the right kind of rational recollection of the process of experience that produced it" (Brandom 2009, 101). Basically, he expresses the thought that there is a communicative form that can describe and guarantee progress for past events and for future events as reality appears through them. Instead, as I will show later, Hegel defends a process of reflection that is related to experience and which depends on the history of how we experience. Hegel is not interested in predicting the future. In contrast, Brandom commits to the reading of Absolute Knowing as ahistorical. Brandom thinks further that such a kind of correct, formal process would bring us closer to mind-external reality:

> The right kind of recollection is one that picks out a trajectory through the previous results of one's actual integrations that is *expressively progressive*. That is, it must exhibit a history that both culminates in one's current view and has the form of the gradual making *explicit* of what can now retrospectively be seen all along to have been *implicit*. Doing that is showing for each previous episode (of those that are selected

> as, as it were, precedential, as revelatory of what one now takes always already to have been there) how that set of commitments can be seen as a partial, and only partially correct revelation of things as they are now known (or at least taken) to be. That is, one must show how each of the recollectively privileged prior integrations made *progress* towards one's current constellation of commitments—both in the judgments that are endorsed and in the consequential and incompatibility relations taken to articulate the concepts applied in those judgments.
>
> BRANDOM 2009, 101

It is not necessarily false to apply this idea of progress to Hegel as long as it is shown as progress with respect to former views and their standards. Yet, according to Brandom, our views are ultimately responding to an unchangeable ahistoric standard of mind-external reality. In this sense, progress means to relate to a mind-external reality in a better way, a claim that Hegel would not defend. For Brandom, history is the process through which we make an implicit, mind-external reality explicit. Brandom further claims that this presupposition of progress is part of our communication:

> In taking one's current commitments as the standard to judge what counts as expressive progress, one *is* taking them as the *reality* of which previous constellations of endorsements were ever more complete and accurate *appearances*. That is the lesson that the normative understanding of the representation relation teaches: what is represented is what serves as a standard for assessing what thereby, in this normative sense, counts as a representing (an appearance) of it.
>
> BRANDOM 2009, 101

In other words, history is for Brandom embedded in his overarching theory of communication that is ahistorically grounded in semantic communication about how the world really is. Thus, he proposes an ultimately ahistoric model that is supposed to secure a better grasp of mind-external reality. Since Hegel's account, however, is historicist, which I think means that all our first-order knowledge claims are limited to the historical moment, Brandom's theory is very different from Hegel. As I articulated in the introduction, this discrepancy positions them as competitors. I believe Hegel's approach is more viable because we, despite Brandom's efforts, currently lack adequate evidence to substantiate the objectivity of our discourses concerning mind-external reality.

Brandom's explicit ahistoricist position is also recognizable in his book *Articulating Reasons*. There, Brandom introduces things that have "natures rather than history" (Brandom 2000a, 27). He is careful insofar as he notes that this distinction between nature and history is itself a "cultural formation" (Brandom 2000a, 27). He evaluates this formation as a "thing that itself has a history rather than a nature" (Brandom 2000a, 27). Brandom concludes that the distinction is unreliable because it is only historical. As Brandom does not delve further into the dialectic between the concepts of nature and history, I find myself compelled to interpret Brandom within a naturalist paradigm—a paradigm that he explicitly wants to avoid. Indeed, he explicitly denies "naturalistic vocabulary", since it cannot explain what we mean by our statements, or how we make statements at all (Brandom 1997a, 154). In this sense, Brandom also acknowledges that "even concepts such as *electron* and *aromatic compound* are the sort of thing that has a history" (Brandom 2000a, 27). Yet, he expounds that "they are not *purely* historical" (Brandom 2000a, 27). Without a discussion of the dialectical relation between these concepts, the distinction between historical and not purely historical remains unclear. Not being 'purely historical', however, implies that there is an underlying mind-external reality guiding us. This mind-external realism is clearly expressed in the following statement:

> For the proprieties governing the application of those concepts depend on what inferences involving them are *correct*, that is, on what *really* follows from what. And that depends on how things are with electrons and aromatic compounds, not just on what judgments and inferences we endorse
>
> BRANDOM 2000a, 27

The inclusion of the word 'really' tends to diminish the significance of other theories that rely on inference and historical formation. It presupposes a foundational element that must possess a nature, ultimately ahistorical, and likely not inferential. Despite this, Brandom does not offer clarification regarding what his realism about electrons specifically entails. However, he presupposes that our concepts must find grounding in mind-external entities that are both real and determinate, implying an ahistorical nature.

To sum up, Brandom suggests that underlying history is something with an ahistorical nature, something determinate and real. However, while denying reality seems absurd, it is also questionable to simply assume a determinate reality as the foundation for our histories and systems of knowledge without sufficient proof. Brandom does not extensively discuss the concept of history; rather, he adheres to a position in which some things have no history. In contrast, Hegel argues that all that we have investigated so far was ultimately within history.

4 Brandom's *A Spirit of Trust* and Conceptual Realism

In his book *A Spirit of Trust* (2019), Brandom presents Hegel's "radically new conception" of conceptual content and explicates the normative implications of Hegel's philosophy (Brandom 2019, 2). There are, according to Brandom, two radically new ideas. First, conceptual content depends on us endorsing or rejecting it with regard to its compatibility with other contents of our thought. Brandom calls this 'modal realism'. Second, Hegel explores the relationship of conceptual content to our normativity. That is, adjusting our normative attitudes to this modal realist understanding of content will eventually pave the way in a new, morally better world. By following modal realism, historical human beings eventually recognize that a final, determinate conception of mind-external reality is impossible. While they are finite subjects immersed in a socio-historical context, mind-external reality is inexhaustible and infinite. Humans can only reconstruct this mind-external reality with the limited, semantic tools of their historical moment. By embracing the modal realist tools and by recognizing the impossibility of a finalizing concept of reality, humans can form a community in which they confess and forgive their inevitable mistakes. Humans recollecting this history of applying semantic tools eventually become aware of their path of error, but by virtue of forgiving on the basis of mutually recognizing their finitude, they gain, at the same time, trust in the history of semantic error-correction. 'A Spirit of Trust' therefore means that we are allowed to commit mistakes and so can correct them as an intersubjective community. Human history is then a progressive series of corrected mistakes toward the ideal of grasping mind-external reality. By mutual recognition and recollection, humans then create a society that makes them better, "magnanimous" people (Brandom 2019, 32). This is why Hegel is the "prophet" of a morally improving world that is guided by and gradually reveals mind-external reality (Brandom 2019, 470). As it should be clear by now, I regard Brandom's ahistorical standard of a mind-external reality as the major problematic premise in his theory. My reconstruction of his book will therefore concentrate on this aspect.

Brandom's work has received much attention within the philosophical community. This book is no different. Major Hegel interpreters, such as Pinkard and Houlgate, but also Žižek have commented on it so far. Houlgate, for example, writes: "Indeed, the book sets out Brandom's principal ideas in a more comprehensive way than even his influential work *Making It Explicit* (1994). It does not replace the latter but builds on it and must now count as Brandom's most significant contribution to contemporary philosophy" (Houlgate 2020). With regard to its attempt of unifying Continental and analytic philosophy, Gleeson calls Brandom's book "groundbreaking" (Gleeson 2020). Yet, he also notes that Brandom remains "a philosopher's philosopher," not attracting attention beyond a small circle of experts (Gleeson 2020).[104] With the "forbidding length" of 800 pages, it is remarkably longer than Hegel's *Phenomenology* (Sachs 2020, 7). Given its dense "writing style" and its technical jargon, Sachs concludes that only a "few will have both sufficient training and sufficient time to read it" (2020, 7). He therefore asks: "for whom is it written" (Sachs 2020, 7).

In addition to the criticism of its small target-audience, Brandom is criticized for neglecting the "accepted standards of scholarship" (Sachs 2020, 3). In that regard, Pinkard classifies the genre of the book as "an instance of … 'here's what I in particular make of large parts of Hegel's book'" (Pinkard 2020, 1). He points out that the book "ignores almost entirely all the other secondary literature on the topic" (Pinkard 2020, 1). With regard to the primary aim of the book—to introduce us to the underlying structure of our normative language that will make us magnanimous people–Houlgate criticizes the specific writing style as not fulfilling Brandom's central demand for mutual recognition (Houlgate 2020). As well as Pinkard, he points out that Brandom "almost completely ignores other commentators" or even "dismisses" them (Houlgate 2020). Moreover, for Houlgate as for many other interpreters, "Brandom's reconstruction of Hegel's phenomenology is deeply problematic" (Houlgate 2020).

Brandom's controversial thoughts on Hegel have contributed to productive disagreements and supported a return of Hegel in analytic philosophy. His creative linguistic interpretation made Hegel's thoughts available to the language of our historical moment. My critique of Brandom focuses less on his omission of other commentators and his appropriation of Hegel to suit his own objectives. The primary concern lies in his focus on mind-independent reality, which, I believe, fundamentally contradicts the Hegelian notion of historicity. It is therefore not a matter of determining the accuracy of Brandom's interpretation of Hegel, but rather assessing whether his approach can effectively rival the framework of historicity found in Hegel's philosophy.

Brandom's book follows three steps. First, he presents his thesis of conceptual realism and reconstructs Hegel's, in his view, semantic perspective on

104 Gleeson, Jules. "Robert Brandom, a Philosopher's Philosopher" *JstorDaily.* 2020. https://daily.jstor.org/robert-brandom-a-philosophers-philosopher/.

conceptual content. Second, he analyzes the relationship between normativity and semantics with regard to Hegel's idea of mutual recognition. Third, he introduces us to objective idealism, including Hegel's understanding of Spirit which, according to Brandom, is the vision of a better, future society. My reconstruction of Brandom's Hegel interpretation will follow the outline of the book: I begin by discussing the conceptual realism that he applies to Hegel. Next, I present his analysis of Hegel's normative understanding of conceptual content. Finally, I will look at the historical perspective that Brandom finds in Hegel's 'edifying' philosophy (Brandom 2019, 32). In my view, Brandom's discussion of history remains a rejection of historicity since it ties history to his semantic concept of a mind-independent and unhistorical world.

4.1 *Brandom's Conceptual Realism Applied to Hegel*

The basis of Brandom's Hegel interpretation is modal realism. Modal realism means here that conceptual content is understood with regard to its compatibility with other content of thought. It is from this semantic basis that Brandom develops conceptual realism. In applying this to Hegel's objective idealism, phenomena are then to be understood as things-for-themselves that stand in relation to things-in-themselves. Things-in-themselves are, in Brandom's terminology, noumena. This transcendental interpretation is controversial. It is in stark contrast to a dialectical interpretation of Hegel, according to which things-in-themselves and things-for-themselves are a mind-dependent construction in order to grasp experience. In my view, it includes also the problematic presupposition that there is a mind-external reality independent of our historical construction that is to some degree knowable. In the next point, I will discuss the problems arising from Brandom's interpretation in detail.

4.1.1 Conceptual Realism

Brandom's "story" of Hegel revolves around the idea of "conceptual contentfulness" (Brandom 2019, 636). This means that things are understood through, first, "relations of [...] 'determinate negation' and, second, through 'mediation' to other such things" (Brandom 2019, 2).

First let us look at 'determinate negation'. Brandom takes Hegel to be a thinker who puts the principle of non-contradiction (PNC) at the heart of his philosophy. In this sense, he translates Hegel's idea of 'determinate negation' as 'material incompatibility'. 'Material incompatibility' refers to the idea that a monochromatic object cannot be, for example, red and green (not-red) in the same respect. It implies that any object is materially incompatible with other objects that display strongly negating features. So, for example, a white cube of salt is incompatible with a cube of iron because of excluding properties.

Material incompatibility therefore fulfills the role of conceptually differentiating a thing from other things. If the criterion of material incompatibility were weakened as a fundamental structural component of objects, all objects would be perceived as logically one. This would be a form of the principle of explosion, according to which our logical apparatus collapses as we contradict the principle of non-contradiction (PNC).

The second aspect we are concerned with is 'mediation', a concept Brandom introduces to amend the strong criterion of determinate negation. In his translation of Hegel's term 'mediation,' Brandom sees it as a form of semantic compatibility. According to Brandom, features of objects not only exclude but must also be compatible with other features. This compatibility is vital for maintaining the unity of a single object, preventing it from disintegrating into multiple objects due to the excluding features. For instance, if an object is red, it can also be round (not red). This mediation of indifferent attributes represents a form of weak negation. It sets those attributes apart from each other, but their difference does not preclude them from being possible features of the same object.

The interplay of determinate negation and mediation, as Brandom argues, ensures the oneness and separability of objects. Both aspects are necessary for explaining the coherence of individual objects. Brandom identifies this justification of the principle of non-contradiction (PNC) as Hegel's modal realism.[105]

Brandom's next step is to reconstruct Hegel's 'conceptual realism'. His reading of the *Phenomenology* is guided by Frege's distinction between sense and reference. Frege's distinction between sense and reference is generally well-known. 'Reference' is the mind-external object that an expression signifies. 'Sense' expresses the way the object is given to the sign-user. Applied to Hegel, things appearing for consciousness are 'sense'. Things thought as in themselves are 'reference'. Fundamental questions emerge of whether Hegel had such a transcendental meaning in mind when he was talking about things-in-themselves [an-sich]. The peculiar reading is the reason why Sachs calls Brandom's Hegel "Fregel" (Sachs 2020, 3). The core problem for Brandom with

105 One may legitimately ask: why is this considered modal realism? Brandom, in contrast to Lewis (who was one of his doctoral advisors), does not endorse the view that possible worlds have objective reality. However, he acknowledges their reality within the context of our practical construction of language. This implies that numerous linguistic constructions of objects become feasible based on the specific semantic network constructed by an individual speaker. Objects are then positioned in relation to their compatibility with this linguistic framework, which can significantly differ from that of other speakers and thus constitute distinct possible worlds.

regard to these two perspectives on objects, however, is the emergence of an unintelligible "gulf separating consciousness from what it is consciousness of" (Brandom 2019, 36). According to Brandom, Hegel argues in the introduction that the "operation of cognitive faculties to bridge it or reunite the two sides" can only "reinstitute the gulf" (Brandom 2019, 37). Bad theories of representation divide sense from reference (see Brandom 2019, 37) and finally cut off the "knowing and acting subjects from the objective world" (Brandom 2019, 82). The challenge, according to Brandom, is to explain how reference and sense can be thought together or how to achieve "categorial homogeneity of the way things are in themselves and the ways they can be for consciousness" (Brandom 2019, 428). Brandom reads Hegel therefore as suggesting an alternative epistemological theory that unifies things for consciousness and how things are in-themselves.

The argument is that Hegelian conceptual realism can avoid the problems of prior representationalist theories (Brandom 2019, 428). The central issue is the "intentional nexus of knowledge" (Brandom 2019, 81), the hidden connection between two ontologically different sides of the object. Former theories fall for a false "fear of truth" as Brandom cites Hegel (see Brandom 2019, 44). Hegel, however, develops a theory that keeps both sides of the object connected. In order to not deepen the gulf with a false theory and relapse into skepticism—the fear of truth—Brandom postulates four conditions that have to be fulfilled for an epistemological theory. First, the "Genuine Knowledge Condition (GKC)" demands that any "epistemological theory" must allow for the "possibility of knowing how things really are" (Brandom 2019, 44). In other words, our knowledge aims at stating the way the world is. I think this criterion is reasonable. Second, the "Intelligibility of Error Condition (IEC)" specifies the GKC. Obviously, not everything that we know can be known immediately. If this were the case, there would be no epistemological problem. Our minds would just have objects without wondering about their ontological status. This implies that error must be possible, which an adequate theory has to explain. Third, the Rational Constraint Condition (RCC) means that anything that is "represented" must serve as a reason for the knower since otherwise knowing would be meaningless in the holistic net of inferences (Brandom 2019, 46). Brandom ties semantics to pragmatism here. It means that knowledge only exists because it has pragmatic significance in the semantic net of concepts. The fourth condition, the "Mode of Presentation Condition (MPC)", is what I regard as most problematic. It means that "appearances must make some reality semantically visible" and thus be clearly related to the way things really are (Brandom 2019, 46). As Brandom boldly claims: any other approach "is doomed" especially if "a psychological conception [...] restricts conceptual content to the subjective side" (Brandom 2019, 81). Falsely assuming that all we know is the mind will, according to Brandom, inevitably lead to the "gulf of intelligibility" (Brandom 2019, 81). Derrida, as Brandom implies, belongs to the group of mistaken philosophers who defend a merely psychological conception. He "offers a picture of a world consisting only of signifiers, with the only things available to be signified being further signifiers" (Brandom 2019, 52). Brandom evaluates that here "things have gone badly wrong" (Brandom 2019, 52). His reading of Derrida, however, may miss the point of Derrida's concept of deconstruction. As far as I understand Derrida, his point is not to say that the world only consists of signifiers. He rather wants to express that meaning constantly shifts, which overall leads to contradictions. The idea that we can obtain a stable picture of the world is therefore hard to maintain. In a sense, Brandom even agrees with Derrida when he later emphasizes that Hegel replaces the categories of *understanding* with categories of *reason*. In this particular case, he elaborates on the idea that our understanding of the world is fluid. In more detail, Brandom argues that while Frege's meaning of senses is stable, for Hegel, the limits of a particular concept can change (see Brandom 2019, 429). In this way, Hegel's concept of knowledge is linked to a constant revision of our concepts. Knowledge is not based on rationality [*Verstand*], which operates with stable concepts and categories, but can be reconstructed as a rational [*Vernunft*] procedure to make our representations of reality more adequate (see Brandom 2019, 431). Of course, Derrida would doubt that such a concept of *Vernunft* serving as a stabilizer for our discourse procedures could be once and for all extracted, while Brandom's point is to distillate such a stabilizer from Hegel's thoughts.

Replacing a mere theory of rationality by a theory of reason ensures that we have genuine knowledge (GKC), but also allows for error (IEC). That is because we do not grasp the world directly, but approach it in a process of correction. Moreover, the produced concepts are used as reasons in social and historical discourses in order to mediate our personal, social, and historical experience thus fulfilling Brandom's rational constraint condition (RCC). However, for his theory to work, Brandom also claims that the world itself must be "conceptual" which means that mind-external objects are governed by "relations of material incompatibility and consequence" (Brandom 2019, 54). Under this condition, "the objective facts and properties natural science reveals as physical reality are themselves in conceptual shape" (Brandom 2019, 54). The major point of Brandom's interpretation of Hegel is this particular mind-external, or in his words, "non-psychological" realism. For Brandom, there is a mediated relationship of our concepts to reality:

> What we talk and think about, what we refer to, what is represented when we apply concepts in judgment and action, is what things are in themselves, how things really are. The contents of our representings, what we express and grasp, what presents those represented things to us, is what things are for us, how they appear.
>
> BRANDOM 2019, 424

However, it remains unclear how to distinguish between the way the world is and the way we take it to be. In that regard, Brandom suggests an unorthodox interpretation of what are 'noumena', 'phenomena', and 'appearances'. For example, he claims that "phenomena are appearances" (see Brandom 2019, 424). He then specifies that phenomena are appearances of noumena that is "what things really are, in themselves" (Brandom 2019, 424). But can we equate phenomena and appearances and relate them to noumena in this way? It seems, for example, that the point of Kant introducing the term 'phenomenon' is to distinguish it from the term 'appearance'. While appearances stand for something, a phenomenon describes something that simply occurs to us without specifying an underlying content. Moreover, the noumenal realm is not simply causing appearances, but is, to speak with Kant, a transcendental necessity. Brandom, however, does not only equate appearances and phenomena, he even equates noumena and phenomena:

> The way things are for consciousness can be just the same as the way they are in themselves. Noumena are a kind of phenomena. They are categorially homogeneous: the same kind of thing.
>
> BRANDOM 2019, 427

In Brandom's view, noumena determine phenomena. Although it is questionable whether we can simply know the thing-in-itself, Brandom suggests that we know how it shines through the veil of phenomena and defines it as the noumenon.

Houlgate, in a charitable reading, remarks that Brandom "claims no direct access to reality" (Houlgate 2020). Rather, he "bases his conceptual realism on what is required for *knowledge* to be intelligible" (Houlgate 2020). In other words, Brandom proposes a transcendentalist reading of Hegel so "that the conceptual form things have for thought *represents* the way they are in themselves" (Houlgate 2020). However, even this transcendental reading is problematic. Pinkard, for example, explicitly rejects such a transcendental interpretation: "In Hegel, there is little to no transcendental search, but there is dialectic" (Pinkard 2014, 16).[106] Dialectic is incompatible with transcendentalism, as it does not allow for a settled relationship between *phenomena* and *noumena* that Brandom suggests.

I understand Hegel differently than Brandom. Hegel is interested in how consciousness experiences things whereas it takes them to be as how things are in themselves. Moreover, in the introduction to the *Phenomenology*, Hegel notes that this in-itself is also a concept that is produced by consciousness. Since, moreover, consciousness is involved in a socio-historical process, the meaning of this 'in-itself' cannot be ultimately clarified. The in-itself is also for consciousness so that the investigation is ultimately about consciousness and its own social and historical constitution. Contrary to Brandom's reading, the question of history is more important than the question of semantics and I think that this is the aspect that makes both theories strict competitors.

Brandom ties Hegel's project closely to a determinate relationship between a mind-independent noumenon and the phenomenon of experience. As he points out, both must be governed by "relations of material incompatibility and material inference" (Brandom 2019, 427). Despite his reading of the terms 'noumenon', 'phenomenon', and 'appearance', his overall argument is, indeed, transcendental: only in this way, "Hegel's conceptual realism" would avoid the aforementioned "unbridgeable gulf of intelligibility" (2019, 427). He then proceeds with the interpretation that "for Hegel, the way the world really is can be said [...] of it" (Brandom 2019, 428). Mind-external reality is accordingly understood as making its appearance in an infinite amount of compatible and incompatible relations. This conceptual realism is what I regard as the major problem of Brandom's interpretation. It stands in contradiction with a pragmatism that takes our historical limitations of recognizing reality into account. As I claim, Hegel's theory is not about semantics but about this question of history. In other words, we need to work around the problem that we do not know mind-independent reality.

4.1.2 Brandom's Hylomorphism and 'Implausible' Idealism

Human community members do not necessarily alter their position in the face of contradiction; instead, the recognition of error is intimately connected to the normative framework of a society and individuals' attitudes and stances toward errors. This observation extends to empirical knowledge, revealing

106 Pinkard, Terry. "Semantic Self-Consciousness: Brandom On Hegel (paper given at a conference): Language and Modernity: Brandom's Semantic Reading of Hegel's Phenomenology of Spirit. (Draft)" 2014. Accessed on June 14, 2021. https://www.academia.edu/7917240/Semantic_Self_Consciousness_Brandom_On_Hegel.

that the reality of external material relations is not simply given and readily accessible. According to Brandom, Hegel teaches us a valuable lesson about how the conceptual content of a mind-independent reality can function as an authoritative guide for beliefs within our communities. In order to develop this account of conceptual realism, Brandom reconstructs the interdependent relationship between norms and semantics that underlies all of our discourses.

However, a significant challenge arises from the fact that while our semantic vocabulary and the resulting insights depend on the norms of our communities, the norms of these communities, in turn, rely on the semantic vocabulary. Therefore, before delving into the normative aspects of modal realism, Brandom must elucidate the intricate interplay between a semantic understanding and a normative understanding of reality. Brandom employs the term "bimodal hylomorphic conceptual realism" to capture and express this interdependence (Brandom 2019, 84).

"Hylomorphism" means that the subjective, normative vocabulary and the semantic, objective vocabulary are both interdependent forms. As Brandom claims, they are both perspectives on an underlying mind-independent reality. For analyzing Hegel's *Phenomenology,* Brandom distinguishes these two perspectives as form and content, or respectively as the subjective perspective and the objective mind-independent world that is thought without subjects. Right at the beginning of *A Spirit of Trust,* Brandom introduces this "hylomorphic structure of content and form" and claims that it supports "conceptual realism" (Brandom 2019, 2). To recall, 'conceptual realism' means that the "objective world" comes "in a conceptual [...] shape" which is independent of subjects (Brandom 2019, 2). For Brandom, it is of utmost importance to defend this independent status of the objective world. In his words, this objective world "does not owe to any activity by the thinking subjects" (Brandom 2019, 3). He emphasizes that "also without subjects there is truth" (Brandom 2019, 84). This true, mind-external world is governed by "alethic modal relations of relative compossibility and necessitation" (Brandom 2019, 84). Brandom's major intent is to first defend Hegel against the falsely attributed and absurd belief of early analytic philosophers, specifically the view that Hegel thought everything is only mind. The mind-independent reality exists, "even in worlds that never included knowing and acting subjects who applied normatively articulated concepts in undertaking and rejecting commitments" (Brandom 2019, 84). For his defense against these interpretations, Brandom utilizes his concepts of reference- and sense-dependence.

In the context of hylomorphism, sense-dependence means that the terms 'world' and 'subject' are only comprehensive in relation to each other. The fact that the concepts of subject and world are related, however, does not imply reference-dependence. With regard to 'implausible' Idealism, this reference-dependence means that the world is dependent on the mind. It means that if the subject disappears, the mind-dependent concept of the world disappears. However, only because we cannot think the world without subjects somehow constructing the concept 'world' does not imply that a world without subjects is impossible. Brandom explains this by way of the "response-dependent concept [of] beauty*" (Brandom 2019, 83). He defines that something is beautiful* if an object causes "a response of pleasure in a […] subject" (Brandom 2019, 83). According to this definition, the sense of beauty* depends on subjects feeling pleasure (see Brandom 2019, 83). Yet, this does not mean that the absence of subjects perceiving beauty means the absence of beautiful* objects. There could still be beautiful* objects, for example, rivers and lakes, that would cause pleasure if subjects perceived them. Beautiful* objects are therefore not reference-dependent on the perceiving subjects. It is only the sense of the objects that depends on the perceiving subjects. The same, according to Brandom, is true for the concept 'mind-external reality'. It is sense-dependent on subjects, but this does not mean that the referent carrying the sense of 'mind-independent reality' does not exist if subjects did not exist. On the contrary, reference-dependence would mean that the world depends entirely on the subject so that if subjects vanished, the world would vanish. Hegel, indeed, does not defend this position.

As Brandom writes, Hegel's idealism is not comparable to "the objectionable Berkeleyan reference-dependence […] subjective idealism" (Brandom 2019, 83/84). Instead, we deal only with the problem of the "amphibiousness of conceptual content" (Brandom 2019, 84). This means that content can be reconstructed in normative and semantic vocabulary, which both depend on each other. However, this interdependence strongly suggests that there is no direct access to a mind-independent reality. When Brandom writes on the conceptual structure of this mind-independent world, he uses a conditional, indicating that modal realism is a possibility for interpreting this mind-independent world:

> The objective world would still be conceptually structured in the sense of consisting of facts about objects and their properties and relations, articulated by alethic modal relations of relative compossibility and necessitation, even in worlds that never included knowing and acting subjects who applied normatively articulated concepts in undertaking and rejecting commitments.
>
> BRANDOM 2019, 83

So far, the argument demonstrates only that sense-dependence does not mean reference-dependence. The concept 'world' depending on the concept 'subject' does not imply that the mind-external world does not exist. Nevertheless, the argument does not show that mind-external reality is modal or conceptual. Although Hegel does not deny the existence of reality, the question remains whether he commits to Brandom's claim that the mind-independent object 'causing' our experience is conceptual. Brandom is adamant about it and states firmly that modal realism is an "essential foundational element of Hegel's idealism" (Brandom 2019, 3). So Brandom wants to relate his hylomorphic structure of form and content to "what are things in themselves" (Brandom 2019, 85). According to him, "we can see how our commitments are intelligible as appearances of an objective reality" (Brandom 2019, 85). He pushes in the direction that commitments are only meaningful if they are the result of a conceptual reality. In other words, modal realism—the way we construct objects—implies conceptual realism. So he writes: "the modal realism that consists in taking laws of nature seriously entails conceptual realism, which entails objective idealism" (Brandom 2019, 84). His argument for conceptual realism can be derived from his proposed hylomorphism in this way:

1. Subjects can think semantically only if the world has an impact on them.
2. The world has an impact on the subject's thoughts, only if subjects represent the world genuinely and can make errors.
3. Only if the world is conceptually structured, subjects can represent the world genuinely and can make errors.
4. Subjects think semantically.

Therefore, the world is conceptually structured.

The argument concludes that the experience underlying reality is conceptual, but this conclusion is still not entirely secured. If my reconstruction of Brandom's argument is correct, then it is circular. 'The world can only have an impact on a subject's thinking if the subject can make errors' and 'Subjects can only make errors if the world has an impact on them' begs the question. Thus, this cannot be an exhaustive argument for the fact of a conceptual reality. Here, it only describes Brandom's "bimodal hylomorphic conceptual realism" (Brandom 2019, 84). According to this model, "the objective and the subjective [are] two different forms that conceptual content can take" (Brandom 2019, 84). They are sense-dependent. For proving his conceptual realism, Brandom therefore introduces additional arguments.

4.1.3 Brandom's Standard of Philosophy—Experiencing Error and Experiencing Reality

As I have presented Brandom's approach, a central intention is to save Hegel from the accusation of being an "implausible" Idealist (Brandom 2019, 90). 'Implausible Idealism' claims, according to Brandom, that there is no reliable grasp of mind-external reality. In order to prove our grasp of reality, Brandom explores the criterion of error (IEC). As he wants to demonstrate, the experience of error is possible only on the basis of a conceptual reality. This thought therefore plays the fundamental role in his "pragmatic metaconcept" of experience (Brandom 2019, 676). Error occurs once the "intentional nexus" of "the subjective realm of thought" and the "objective realm of being" fall apart. It is the result of a comparison between "what things are explicitly 'for consciousness' and what things are "in themselves" [an sich] ('implicitly')" (Brandom 2019, 666). Error according to this conception will lead us closer to what reality really is, even though we will never fully grasp it.

In regard to his theory of error, Brandom notices first of all that Hegel makes some "very odd" claims (Brandom 2019, 89): if our assessment of an object changes, the object itself changes. Brandom wonders "why should we think that, when a commitment a subject took to express how things really are [...] is revealed as expressing merely how things are for consciousness, the reality changes?" Hegel makes this claim in §85 of the *Phenomenology*. As already discussed, Brandom reads Hegel's *Phenomenology* through the lens of Frege's distinction of sense and reference (see Brandom 2019, 666). This removes a central idealist characteristic from Hegel's philosophy, i.e., the thought that the in-itself is something that is for consciousness. Instead of working through the dialectic of experience, Brandom now reconstructs subjects as adapting their beliefs to that which is revealed as mind-external reality. Once mind-external reality shines through the veil of our limited phenomenal perspective, we see that our cognition was just an appearance or a phenomenon. As also already discussed, Brandom does not distinguish the terms 'appearance' and 'phenomenon' sharply, so that, according to him, the noumenal "reality" behind a phenomenon can be "unmasked" (Brandom 2019, 677). But how can we say that we have really grasped reality through the experience of error?

In order to explain the experience of error in greater detail, Brandom refers to the common example of a stick that appears bent while submerged partially in water, yet appears straight when it is pulled out of water. His interpretation of this example reveals a rather simplistic understanding of what phenomena are: "Pulling [the stick ...] out of the water, one sees that it was really straight all along" (Brandom 2019, 677). Somehow we obtain the ability to differentiate: "how things merely look from how they really are" (Brandom 2019, 677).

This is how Brandom transforms the dialectical question of phenomenology into a metaphysical question of reality. In other words, he transforms Hegel's dialectical question of how consciousness produces the in-itself and the for-itself, into a question of how mind-external reality reveals itself. The stick in the water expresses the appearance of the stick while the unbent stick exhibits more truthfully the stick of a mind-external reality (see Brandom 2019, 89). Yet, how can we claim that we have now reached the moment at which we are better able to recognize how things really are?

A common objection is that the new revelation of the stick is also only a further phenomenon. From here, we could make the argument that the straight stick constitutes the more robust experience because, for example, it is validated through other reasonable tests. We could also argue that our common experience of straight sticks outweighs our experiences of bent sticks. On the basis of such warranting factors, we could come to the agreement that the straight stick is our true standard within our current community. This agreement would not be entirely certain, but it represents a reasonable pragmatic option. Brandom, however, follows a different line of thought.

Brandom critiques an interpretation of Hegel that posits not only a shift in our perspective but also a transformation in the very reality of the object itself. Hegel would turn into an "implausible Idealist", if we misunderstood "the claim he is making" (Brandom 2019, 90). So, first, Brandom demonstrates that in Hegel's view, only our perspective of the object changes:

> What object (here, crucially, in the sense of what conceptual content) the status attaches to is essential to its being that knowing. Altering the knowing, by endorsing a different, incompatible content, alters the status of the original content, and so alters the "object" associated with the original knowing: its status changes from being a conceptual content that is endorsed to being one that is rejected.
>
> BRANDOM 2019, 92

Brandom points out that the subject holds only knowledge of how the object occurs to consciousness. Of course, this occurring knowledge can change from "the stick as bent to being the stick as straight" (Brandom 2019, 92). In this case, the subject, indeed, endorses a different object. So Brandom reduces it to the question of how the "standard change[s] when I realize that my shape-commitment does not measure up to the standard, that it gets things wrong" (Brandom 2019, 89). So far, this is reasonable. Hegel does not claim that reality itself changes. Yet, although Hegel does not deny a mind-independent reality, he also does not believe that we simply progress towards it. Brandom neglects

Hegel's emphasis on consciousness creating its own standard of knowledge independent of a mind-external reality. Consciousness operates on the basis of its experiences, compares and reflects on them. Instead, Brandom attempts to correct Hegel's "extravagant and perverse" claim that "appears to trade on an obviously unwarranted slide" (Brandom 2019, 89). He wants to show why the straight stick is closer to reality than the bent stick. For this, Brandom draws on the meaning of "essential", which for him is not "a symmetrical relation" (Brandom 2019, 89). He illustrates this with regard to one's parents. It is not essential that parents have children, but it is essential that the offspring has parents. Brandom states: "essentiality of origin of humans does not entail essentiality of offspring" (Brandom 2019, 89). He argues further that being bent is not an essential property of the stick. In reality, the essence of the stick is straight. In a sense, he says that the straight stick is the 'father experience', while the bent stick is its 'offspring'. A bent stick cannot produce straight sticks, but straight sticks can produce bent sticks. So, the stick in the water expresses the appearance of the stick, while the unbent stick represents something that is closer to a mind-external reality. But why is the unbent stick the more real stick? Brandom adds that being bent would not be an "essential feature of the stick (the reality)" (Brandom 2019, 89). Yet, is it essential that sticks are unbent?

Certainly, we could introduce some arguments for why bent is a mode of straight, but this seems to be far from certain. Sticks are, moreover, simple examples, but what about the essence of more complex phenomena like atoms? We could also ask: why is it that we talk about representations of sticks in the first place? If, for example, all essential features of the stick depended on our construction of them, then it is only a phenomenon that we discuss. Such a phenomenal consciousness does not have to involve itself in statements about a theoretical mind-independent reality of the essence of straight sticks. It investigates what is presented to it within experience. We could easily understand this as a dialectical, empirical realism. If I am not mistaken, Brandom reveals here some kind of essentialism which is a form of metaphysical realism.

The overall argument of Brandom seems to be that the more appearances we unmask, the closer we get to a stable mind-independent reality (see Brandom 2019, 101). For this reason, Brandom's example of the bent stick is just about restoring reality. We repair our views and bring them "in line with the reality, what things are in-themselves" (Brandom 2019, 88). Further below, Brandom claims that it is about "unmasking of something as not real, but as mere appearance", which eventually eliminates "the specter of skepticism" (Brandom 2019, 97). He writes explicitly on how we explore mind-external reality:

> This process of weighing the credentials of competing commitments to determine which should be retained and which altered so as to remove local material incompatibilities is the process by which we find out (more about) how things really are.
>
> BRANDOM 2019, 101

To reiterate, my claim is that this process of knowledge production regarding "how things really are" which we find in Brandom's *A Spirit of Trust*, is not a Hegelian dialectical reconstruction of our experience, which I take to be central to any "reading" of Hegel. In Hegel's dialectical model, we preserve former moments and involve them in the complex mechanics of negation. In Brandom's model, it is more an approximation to mind-external reality that eliminates former 'false' theories. With a 'Fregelian' semantics then, we deal with appearances that "are found to be materially incompatible with other commitments in the experience of error" (Brandom 2019, 101). And as Brandom repeats over and over: it "is the process by which we find out (more about) how things really are" (Brandom 2019, 101). It is this kind of approach that I find deeply in tension with what Hegel is suggesting and where I want to argue that Hegel's theory is better than Brandom's.

As we can see, Brandom keeps things-in-themselves as an essential moment in his theory. For Brandom, things in themselves are necessary for knowing (see Sachs 2020, 3). Sachs writes with regard to this: "We would not be entitled to regard ourselves as having determinate thoughts at all—thoughts with content or sense (Sinn)—if we could not have any cognitive awareness or grip on determinate things in themselves" (Sachs 2020, 3). As I will show in Chapter Four of this book, Hegel has a different epistemological account in mind that treats things-in-themselves not as mind-independent noumena but as necessary projections of consciousness. This does not mean that Hegel denies reality. He, however, claims, contrary to Brandom, that knowledge is possible without having a secured grip on mind-external reality.

Overall, Brandom's conceptual realism and his interpretation of Hegel hinges on the question of how to understand things-in-themselves. Brandom suggests a transcendental, metaphysically realist, and semantic reading of Hegel's philosophy that I see in conflict with what I take to be Hegel's historicist dialectics, which I regard as the better epistemological approach. I will explain this in Chapter Four in more detail. Now, we move to assess how Brandom, on the basis of his semantics, will develop his normative interpretation of communities. We will discuss communities whose overall goal it is to get closer to mind-independent reality and which, according to Brandom, must structure themselves with respect to the underlying conceptual reality.

4.2 *The Normative Side*

In line with his pragmatist approach, Brandom contends that achieving satisfactory semantics necessitates reevaluating subjective normativity. The subject is confronted with the authority of the objective world (Brandom 2019, 462). However, the acknowledgment of this authority is contingent upon the collective decisions of the community of subjects. Therefore, constructing an appropriate semantics requires considering the role of the community. With regard to this normative semantics, Brandom points to an important difference between objective facts and subjective statuses. Whereas "relations of incompatibility and consequence [...] hold between objective states," they do not "hold between subjective acts of thinking" in the same way (Brandom 2019, 3). For objects of thought, "one ought not to", as Brandom writes, "endorse" incompatible "judgments", or "endorse a claim but not to acknowledge its necessary consequences" (Brandom 2019, 3). Knowing content means acknowledging that perceived objects stand in similar relations as our thoughts stand in relation to each other (see Brandom 2019, 2). It means that in order to gain knowledge, a subject must follow rules which, as Brandom argues, copy the regularities of mind-external reality.

Now, although the rules must be in congruence with mind-external reality, they are not causally enforced. Rather, they are socially mediated. In a word, the world itself is not simply creating our mind. From this follows that a subject does not simply know content. Instead, it must follow social rules in order to institute knowledge. According to Brandom's model, the 'ought' is mediated by social procedures that demand "recognitive attitudes of reciprocal confession and recollective forgiveness" (Brandom 2019, 615).

So far we can see that Brandom's account is not limited to the semantic side of what he calls Hegel's 'ground-level' empirical concepts, but involves how pragmatic normativity shapes these contents. The core idea is that we will develop explicit semantics by deciphering how conceptual reality is mediated through normative behavior. Moreover, this semantics will serve as a guide for our normative communities. Brandom does not only want to explore the grounding elements of our social interaction. He wants to define the requirements "for genuine bindingness of discursive norms" (Brandom 2019, 31).

As his conceptual realism builds the foundation for his Hegel-interpretation, Brandom grasps the essence of Hegel's book as "a semantics with an edifying intent" (Brandom 2019, 32). Following Brandom's model, a future "discursive activity" will make "us [...] better people" or, in more glorious words, we will become "magnanimous people, who live and move and have our being in the normative space of Geist in the postmodern form of trust" (Brandom 2019, 32). The question emerges: what is the normative rule that we have to live by?

"The big Brandomian-pragmatist claim" as Pinkard points out, "is that normative statuses are instituted by normative attitudes, but normative statuses are not themselves just a matter of normative attitudes" (Pinkard 2020, 3). As previously argued, normative statuses manifest in two ways. Firstly, they convey authority, indicating that someone is authorized to make claims. Secondly, they entail responsibilities, signifying that someone is obligated to adhere to the consequences of a claim (see Sachs 2020, 3). Normative attitudes encompass the recognition of authority and the ascription of responsibilities (see Sachs 2020, 3). Consequently, normative statuses and normative attitudes are interdependent and rely on each other for their comprehension. Normative attitudes, moreover, include the acknowledgment of authority and the attribution of responsibilities (see Sachs 2020, 3). As a consequence, normative statuses and normative attitudes are inherently interconnected and reliant on context. This implies that we cannot fully comprehend either of them in isolation.

To illustrate the interconnectedness of attitudes and responsibilities, Brandom draws upon Hegel's master-slave distinction. This conceptual framework portrays the struggle between two consciousnesses, each yearning for independence, which ultimately culminates in a life-or-death battle where both subjects are willing to sacrifice their lives for autonomy. However, one subject chooses self-preservation, leading to the establishment of a deeply distorted relationship between master and slave. In this imbalanced dynamic, the master claims absolute independence as the victor, yet as Brandom's interpretation of Hegel elucidates, "there is no independence without a correlative dependence, and vice versa" (Brandom 2019, 306). Unbeknownst to the master, they remain dependent on the slave. Brandom asserts that this dependency reveals that authority cannot exist without responsibility (Sachs 2020, 4). He argues that the master's perceived independence stems from the erroneous belief that their subjective attitudes establish normative statuses autonomously, disregarding the slave's necessary acknowledgment. During the struggle towards death, the master compels the slave to conform to their viewpoint. In this sense, the "mastery's ideology of pure independence" as a "form of cognitive consciousness" declares itself to be always correct (Pinkard 2020, 4).

For interpreting the master-slave-relationship, Brandom applies his Intelligibility of Error Condition (IEC) and remarks that authority can only be reasonably exercised if there is "room for error" (Brandom 2019, 348). One's attitude cannot simply dictate any normative status since otherwise, any perspective would be correct. Authority must be grounded in some kind of competence and objective reality. The subject alone is not sufficient for defining normative authority but must take external constraints into account. Only a community of subjects who correctly identify each other's powers and who relate to an objective reality implement reasonable authority. Such an authority involves the "others to hold one responsible" (Brandom 2019, 306). In this sense, authority and responsibility are sense-dependent terms.

Brandom emphasizes that his model avoids the pitfalls of a "consensus theory" (see Brandom 2019, 306). In this model, content is not deemed true simply because the master declares it as such and the slaves comply. Truth is not contingent on enforced agreement or solely mediated by social factors. Rather, content is considered true when a community establishes rules based on objective criteria, which in turn carry accompanying responsibilities. The game of giving and asking for reasons is structured in accordance with the objective demands of a community that exists within a shared objective reality. The community orients itself around this content. Those who do not participate in this game risk the decline of the community and, consequently, lose the community's recognition of their authority.

This "reciprocal sense-dependence" of master and slave is what Brandom sees as the core of "objective idealism" (Brandom 2019, 462). The master-slave model and its shortcomings build the ground for Brandom's historical analysis of how an ideal society should be structured. Through historical investigation and repair of former imbalances, we will eventually arrive at the "final, fully developed historical-recollective form of reciprocal recognition" (Brandom 2019, 453). This final form of recognition will be developed by Hegel in the *Spirit* section (see Brandom 2019, 453). In regard to this, Brandom specifies the major goal:

> The principal aim of the book is to articulate, work out, and apply a way of understanding the transition from premodern to modern social practices, institutions, selves, and their immanent forms of understanding.
>
> BRANDOM 2019, 469

The self-understanding of these "social-recognitive and historical-recollective dimensions" will lead to the final form of knowing, i.e., Hegel's 'absolute knowing'. For Brandom, this "postmodern [...] self-conscious form of normativity" can be understood if we study the historical occurrences of "forgiveness, confession, and trust" (Brandom 2019, 464). In the following point, I will discuss Brandom's view of these different forms of society and how they then lead us to the insights necessary for Brandom's age of trust.

4.3 *The Historical Progression—The Advent of a Better Society*

During the discussion of the historical dimensions of his conceptual realism, Brandom reintroduces and specifies the historically dependent categories of reason [*Vernunft*]. While understanding [*Verstand*] confines concepts to one definition that is independent of its historical development, reason is "dynamic" and its concepts can change according to the socio-historical context (Sachs 2020, 6). This is because "the content of a concept" cannot be simply defined independently of its production (see Brandom 2019, 6). Concepts of this dynamic category of reason [*Vernunft*] can be reconstructed in two ways: on the one hand, we can observe retrospectively "what the content has been all along: what norm implicitly governed […] the concept in judgment" (Brandom 2019, 6). This, according to Brandom's before-mentioned, pragmatic principle of mastery, involves that we find out how we have successively improved our grasp of mind-external reality throughout history. On the other hand, we can extend this successful past application prospectively into the future. Then, we correct "partially indeterminate content" and improve it "by applying the concept in novel circumstances" (Brandom 2019, 7). *Reason* is the process through which we master our grasp of the empirical and ethical world by making our concepts more accurate with regard to a mind-independent reality.

The third part of Brandom's book discusses the concrete normative rules that are necessary for the path of improvement towards a better society. Given his pragmatic-semantic model, the formal insights of his conceptual realism must be replicable with regard to the normative aspects of concept application. With regard to empirical content, the claim is that concepts become more accurate in representing mind-external reality. With respect to the ethical world, the claim is that concepts become more accurate with respect to their objective normative application. For his project, Brandom wants to show how both sides, the normative and the semantic side, work hand in hand, whereas, however, the idea of a mind-independent reality is key.

The major idea of Brandom is that society on its path to approach mind-external reality has been transforming towards the ideal of reciprocal recognition, which entails a more and more successful application of concepts. The ideal of reciprocal recognition is, according to Brandom, expressed in Hegel's concept of *Sittlichkeit*—a form of ethical living. *Sittlichkeit* runs through different historical stages of immediacy, mediation, and will ultimately arrive at a mediated immediacy that is a perfected form of reciprocal recognition. The final postmodern *Sittlichkeit* stands for a future society in which individuals live and discuss according to the standard of what mind-external reality 'really' is.

Given his vision of such a future society, Brandom perceives Hegel's message as explicitly edifying (see Brandom 2019, 32), an interpretation that is criticized since Hegel explicitly denied edification (see Hegel, PoS §9, 6). Reflecting on Hegel's last thoughts on the "owl of Minerva", Žižek reminds us that we "*retrospectively* translate[.], into a 'gray' (lifeless) conceptual scheme, a form of life which has already reached its peak and has entered its decline" (Žižek 2020).[107] In other words, there is nothing that warrants the application of our past observations to the future. Since Hegel therefore is "the philosopher most open to the future" Žižek suggests interpreting Hegel's philosophy as "a Spirit of Distrust" (Žižek 2020). He claims that overly optimistic interpreters such as Brandom should be "rejected" (Žižek 2020). For the same reason, he calls Brandom the most prominent "not-yet-Hegelian" (Žižek 2020).

On the contrary to Žižek and in light of Hegel's 'prophecy', Brandom places strong emphasis on "the One Great Event in human history" that initiated the new age and which will lead to "new, better kinds of institutions, practices, and self-conscious selves" (Brandom 2019, 470/471).[108] The 'One Great Event' concerns the transformation of norms from antiquity to modernity that still poses a challenge for our society in which we want to develop a new form of *Sittlichkeit* as a perfected reciprocal recognition. There are, according to Brandom, two problems that we still have to overcome. First, it would be a step back into antiquity, if we understood norms as objective and independent of subjects (see Pinkard 2014, 2). This is what Brandom calls "hyper-objectivity" which is still expressed in some of our moral attitudes. Second, the opposite extreme would occur if we understood norms as constructed only by subjects. This would be "hyper-subjectivity". Both moral theories of norms are insufficient, either underestimating the role of subjective normative attitudes or underestimating the role of objective, normative facts. Brandom seeks a balanced position between normative attitudes and normative statuses. This balanced position, which is his vision of reciprocal recognition, is achieved only if the "agent's authority (as the one doing the binding) is exercised in undertaking a coordinate responsibility (what one binds oneself by)" (Brandom 2019, 462). In light of this, we should understand Brandom's overall reading of the

107 Žižek, Slavoj. "Hegel On The Future, Hegel In The Future." *Philosophy Now* 140, 2020 https://philosophynow.org/issues/140/Hegel_On_The_Future_Hegel_In_The_Future.

108 As Pinkard points out, Brandom's expressions regarding this "momentous" shift "are themselves extravagant" (Pinkard 2020, 5). It is three times called titanic, and once cataclysmic. Moreover, "it forms an 'unprecedented revolution in human institutions and consciousness'" (Pinkard 2020, 5).

Phenomenology as an allegory of how our history has been progressing towards this ideal of reciprocal recognition (see Pinkard, 2014, 3).

So how do we achieve the balance that Brandom has in mind? The new standard that Hegel recognizes is that the subject, by the means of society, grasps itself adequately as its own object. This identity can be called Spirit. As Pinkard expresses it well, Brandom discusses this identity as "a kind of inferentialist semantics grounded in a normative pragmatics" (Pinkard 2020, 1).

Spirit in its most genuine form—a normative pragmatics guided by conceptual realism—is not just there. The identity of the subject and object has to be achieved through a history of forgiving and confessing errors. The three major stages of this history of errors are according to Brandom: first, the ancient society of 'God-given' norms, second, the modern society of norms and truths relativizing individuals, and finally Brandom's postmodern society that synthesizes the former two extremes. Concretely, this envisioned post-postmodern society expels moral and alethic relativism but also denies that norms can be truly independent of the subject or that we have a direct grasp of mind-independent reality.

The first of these stages is the ancient society with a "hyperobjective traditional picture of normativity" (Brandom 2019, 512). Back then, norms were understood as independent of normative attitudes. Accordingly, people did not base normativity on recognition. Instead, norms were seen as externally given as if they were made by Gods. For the same reason, people did not distinguish between deed and action (see Brandom 2019, 387). The outer appearance of the action, i.e., the deed, was evaluated without considering the subject's intention. Oedipus' deeds, for example, were evaluated according to their outer appearance, so that despite him acting genuinely with regard to his individual perception, his deeds were evaluated as a moral failure. He failed with regard to the hyper-objective norms of his historical moment.

With reference to Antigone who wants to bury her brother, but stands in conflict with the fact that he has betrayed the state, Brandom demonstrates how such hyper-objective norms ultimately lead to contradictions. Determined once and for all, their finite character will lead to incompatible commitments of the involved subjects.

The insights taken from these allegories prepare the transition from immediate *Sittlichkeit*—norms are just what they are—towards the second stage of mediated *Sittlichkeit.* In modernity, norms are understood as the result of subjective attitudes. The reversal of this understanding is what Brandom calls 'the One Great Event'. The subject discovers that commitment to norms always involves a choice to acknowledge them. Thus, it discovers its autonomy in determining norms. However, this immediately leads to another extreme, the idea that subjects alone determine the content of norms. In the aftermath of the 'One Great Event', norms lose their objective status. Brandom summarizes the major problem of this position: "But if there really are no norms, then the attitudes themselves can have no real content" (Brandom 2019, 512). In other words, extreme moral relativism, according to Brandom, leads to the impossibility of moral rules.

Although subjects in modern societies obtain freedom, they become alienated from their societies that have provided norms on historical, objective grounds. Using Brandom's vocabulary, modernity reveals the sense-dependence of normative statuses but misunderstands it as reference-dependence. Subjects believe that only their normative attitudes matter and thus subjects lose their trust in societies. Thus, according to Brandom's arguments, modernity enters the age of ungrounded relativism, close to rejecting any genuine content.

The challenge is now to mediate both forms of society and to achieve a new form of mediated, immediate *Sittlichkeit* that Hegel, according to Brandom, calls "Absolute Knowing" (Brandom 2019, 472). This will be the third stage that Brandom announces as the age of trust.

Hyper-subjectivity and hyper-objectivity are still at work at our current historical moment and the opposition can be displayed in two attitudes of how we interpret actions. Brandom distinguishes, according to Hegel, the base [*niederträchtig*] and magnanimous [*edelmütig*] attitude. His distinction is based on Hegel's interpretation of the saying that "no man is a hero to his valet; not, however, because the man is not a hero, but because the valet—is a valet" (Brandom 2019, 561). A hero regards norms as having value in themselves and, moreover, sees it as a duty to follow them. However, while the hero's actions result from seemingly external norms, the valet sees them only as results of the hero's own attitude. The valet, adopting a base attitude, regards the hero as a man who follows norms for his own advantage.

Both attitudes display the opposition we deal with: either normative statuses are the result of self-interest, or normative attitudes are chosen because we follow objective, external norms. Yet, the reason why the hero appears to the valet as a base and selfish individual is not because every hero has quite human characteristics. Rather, it is because the valet cannot interpret the hero graciously (see Pinkard 2020, 4). The valet reconstructs norms from a 'base' perspective. Meaning that all norms from such a perspective are only followed because of self-interest. Interpreting societies on the ground of the base attitude means that any society must be reconstructed as inherently egotistical.

For Brandom, it is a question of how to overcome the base view. His guiding thought is that "the edelmütig attitude as a practical-recognitive commitment" is a transcendental condition for "semantically contentful cognition and agency" (Brandom 2019, 581). Without the higher magnanimous attitude, no norm could be reliably grounded in a society. While relativist modernity would eventually destroy itself, post-postmodern societies that adopt the magnanimous attitude will eventually reach the "final form of Geist, in which normativity has the form of trust" (Brandom 2019, 582). Magnanimity will bring us to the "self-consciousness" that "Hegel calls 'Absolute Knowing'" (Brandom 2019, 581). 'Absolute knowing' is therefore the "practical transformation of self-conscious selves" through which communities recognize themselves in history (Brandom 2019, 582). How do communities achieve this insight into their own historical doing? It is based on a dialectic process, which Brandom explains through the dialectic between "evil consciousness" and a "hard-hearted judge" (Brandom 2019, 584). As a result of this dialectic, the judge will finally drop his base attitude and embrace a higher magnanimous attitude (Brandom 2019, 584).

The task of a judge is to evaluate the individual who, from the perspective of a universal, moral principle, has committed a crime and thus is seen as evil. However, without reflecting on herself, the judge judges the individual only according to the individual's self-interest and compares it to an impossible ideal. In other words, the judge operates on the grounds of a base attitude. Just like the valet who cannot recognize the hero, the judge cannot recognize the individual beyond his, as Brandom frames it, naturalist perspective. She thus has a base stance towards the criminal, "rather than simply acknowledging a fact" (Brandom 2019, 596). Yet, by eventually recognizing herself, she overcomes the base attitude by which she formerly judged the individual and separated the criminal from a community of humans. By overcoming this attitude, she "produces a community" with the judged and moves in the direction of an "edelmütig recognitive" stance (Brandom 2019, 597). Overall, the hard heart of the judge breaks and leads in turn to a confession: her ability to judge was limited.

The judge must confess that her judgment was "infected with the aspect of individuality" (Brandom 2019, 601). The "assessor" has to admit "his subjective inability" to forgive because she does not know things-in-themselves (Brandom 2019, 601). Here, Brandom turns Hegel's ideas into a version of his conceptual realism: the judge judges according to "what the content of the concept was for him, rather than just what it was in itself, an appearance of the reality, rather than the reality itself" (Brandom 2019, 602). In other words, the judge cannot judge entirely according to the universal because she does not know reality immediately. Nevertheless, according to Brandom's model, we will know reality better by virtue of historical mediation. The solution lies in the idea that we constantly improve our knowledge of reality along the traditional line of judges throughout history. “A later judge,” as Brandom argues, will “forgive the earlier judge” by integrating the decision through a “retrospective rational reconstruction of the tradition” (Brandom 2019, 603).

The standard for this long line of judges will be a methodological approximation to “what it really is” (Brandom 2019, 603). In other words, our uniform application of concepts throughout history will bring reality more and more to light (see Brandom 2019, 603). Brandom calls this method a “reconstructive recollection” that is expressively progressive, while “experience of incompatibility is [...] the engine” of the whole process (Brandom 2019, 603). The relationship between the judge and the criminal is only an allegory for this process.

In order to fully explain his argument for this progression towards reality, Brandom now applies his Fregelian distinction of reference and sense again. Overall, the “rational reconstruction focuses on the identity of (Hegelian) reference—the underlying conceptual norm—[...], rather than the disparity between the elements of the sequence of (Hegelian) senses” (Brandom 2019, 603). So, all our actions are guided by the underlying noumenon that the different senses only distract us from. The “noumenon” can be “recollectively found” and unifies the “sequence of disparate phenomena” (Brandom 2019, 603). Brandom also calls this process “recollective labor” (Brandom 2019, 610). In particular, it means that a magnanimous “forgiver” orders “contingency”, for example, the crime of the individual, according to “the form of necessity” (Brandom 2019, 610). It means that the individual who has committed crimes is seen from the perspective of a universal history that has eliminated contingency. However, Brandom notes that ‘necessity’ does not mean that history had to be that way. Rather, it means that all events in history have to be necessarily ordered so that we can gain insight. Part of this necessity is that earlier stages need to be in place so that later stages can built upon them. In other words, “judgments and actions that resulted from particular contingent circumstances” are ordered by a forgiving consciousness so that the underlying reality—in this particular case, it is our morally correct behavior—finally emerges as a result of these necessary stages (see Brandom 2019, 610). Overall, Brandom reads the dialectic of judges as another allegory that stands for how we make progress towards mind-independent reality bringing us to an improved normative semantics. If we do not follow this model, as Brandom intends to warn us, we will fall back into the relativist dilemma of modernity and postmodernity.

Becoming aware of this normative relationship to reality, Brandom sees the “semantic self-consciousness” emerging from the process of recollection. It is “awareness of the transcendental conditions of the intelligibility of determinately contentful attitudes” (Brandom 2019, 615). So we have to explicitly acknowledge the "always already implicit commitment to adopt generous recognitive attitudes of reciprocal confession and recollective forgiveness" (Brandom 2019, 615). It means that the necessary condition for recognizing reality in the long run and living a moral life depend on a forgiving community. We forgive the inevitable mistakes that occur due to our contingent situatedness in a universal history of discovering what reality really is. The 'ought' for why we should not uphold incompatible commitments is grounded in this community that is capable of confessing and forgiving false beliefs and thus makes by virtue of a tradition progression towards reality possible. It makes itself possible as an ethical society of 'magnanimous' humans. Without us constituting such a trustful community, our recognition of reality would simply not progress and our community would fall apart.

Brandom underlines that the magnanimous attitude is not an option but the foundation of a society of trust:

> Practicing the recollective recognitive hermeneutics of magnanimity is not just one option among others. A proper understanding of ourselves as discursive creatures obliges us to institute a community in which reciprocal recognition takes the form of forgiving recollection: a community bound by and built on trust
>
> BRANDOM 2019, 635

An ethical community needs to organize its discourses as a way to accomplish its final criterion for being a community: progressing towards mind-external reality.

So far I have carved out how much Brandom's approach is tied to the concept of mind-external reality. A community of humans builds a normative relationship to mind-external reality and shall not understand it in any other way than through its normative attitudes. Moreover, for a functioning normative community that progresses toward reality, the attitude of forgiving is key. Žižek criticizes that this "recollective forgiving remains an ambiguous notion" (Žižek 2020). He points to the limit of forgiving which involves the question of whether we, for example, can forgive Hitler (Žižek 2020). Moreover, he asks for the evidence that warrants such a leap of trust. Or as he disparagingly calls it "this faith in progress" (Žižek 2020). Žižek points out that the faithful forgiving spirit is "totally unwarranted, and at odds with Hegel's basic metaphysical stance" (Žižek 2020). Instead, the future is fundamentally open. He even believes that the "Third Phase" points at "new, unpredictable antagonisms and forms of violence" (Žižek 2020). Overall, Žižek warns of the dangers lurking in the optimism of Brandom's interpretation:

> We want to overcome the gap between a society's morality and a free subjectivity which no longer recognizes the morality as its own; but instead of bringing them together in a kind of higher synthetic unity, *we lose both*. For instance, did Stalin not promise to implement a synthesis between a strong communal spirit and free individuality, promising actual freedom? And wasn't the result the loss of freedom itself, in conditions of total alienation?.
>
> ŽIŽEK 2020

Žižek puts his finger into the wound of Brandom's formalist approach. He employs the history of violence that emerged after Hegel and thus discusses the implications of Hegel's dialectical model applied to the 'real' experienced life-world. Given this history of violence, there is no clear indicator for optimism. Were the many victims of the World Wars only necessary mistakes in recognizing the reality on our way of becoming better, 'magnanimous' people? How many humans have to be sacrificed on the path of 'inevitable mistakes'? With Brandom's account, questions of later critical thinkers and postmodernists, such as Adorno or Foucault, remain therefore unanswered: How can we forgive and trust in a society that, at the same time, has committed the most severe crimes?

Of course, Brandom's book has already 800 pages and we cannot expect that all our questions are answered. Nevertheless, as a philosopher, who gives advice for how we should live, he probably needs to indicate how his model responds to the currently demanding challenges of our historical moment. After Hegel, philosophy has made remarkable turns towards practical questions, such as critical race theory, environmentalism or feminism. They are all results of the preceding, brutal history of violence. Can the questions that were raised by history be answered with Brandom's standard of reality and normative progress?

My major criticism, however, is not the lack of application to the current development of practical philosophy. I am also not objecting that Brandom applies his own "terminological apparatus" in order to reconstruct his version of Hegel (Brandom 2019, 307). Brandom's book is an optimistic contribution to the future of our community and how to read Hegel through the perspective of analytic philosophy. Instead, my major criticism is directed towards Brandom's particular mind-external realism that he, from my perspective, falsely applies to Hegel. In my view, Hegel makes a competitive and radical suggestion of how we have to exercise a historicist philosophy that remains open for the future. The metaphysical mind-external standard of reality is replaced by a mind-internal standard of experienced, dialectical truth. It is a reflection on the possibilities of philosophy with respect to the finite, historical human beings we are. In this sense, I will defend a rather historical reading of Hegel that in its epistemology is anti-semantic. Although Brandom invokes a notion of history in *A Spirit of Trust*, I do not believe that it matches Hegel's more productive, relativist notion of history that I will present further below. Brandom's overall approach must be seen as only semantic and thus as opposed to historicism.

In this regard, Pinkard also criticizes Brandom's focus on mind-external reality and is in sympathy with the view of a historical Hegel. In contrast to Brandom's theoretical focus on mind-external reality, Pinkard suggests that concepts "are developed in practice", while "the speculative realm", which may include questions of mind-external reality, helps us to make explicit our own doing (Pinkard 2020, 15). Nevertheless, the speculative realm does not describe the reality of our interactions without qualification. For example, the speculative concept "of the object 'in itself' is a result of history" and therefore also subject to possible changes (see Pinkard 2020, 15). For our historical practice, concepts do therefore not relate to a mind-external reality but rather to "the essence of agency itself" and "are developed in history" (Pinkard 2020, 15). Pinkard specifies this historical process of conceptual development: during concept application, contradictions of insufficient concepts become more and more pressing, until the concepts will change. But this does not mean that we came closer to mind-external reality. Pinkard objects that although Brandom "admits conceptual change," he only speaks about "empirical difficulties" that are always tied to the question of mind-external reality (Pinkard 2020, 15). But, as Pinkard points beyond mere empiricism, the "existent public criteria for possessing a concept [...] do not exhaust its meaning" (Pinkard 2020, 15). In other words, Brandom's model is built upon an inference of mind-external reality that neglects the socio-historical involvement of subjects who develop spiritual communities independently of the question of a mind-external reality. Our take on mind-external reality turns out to be mediated by our socio-historical perspective. We constantly adjust our views, but as Pinkard remarks in that regard: "Changes in the 'speculative' concepts do not make explicit what was already there" (Pinkard 2020, 15). On the basis of this, he further criticizes that, according to "Brandom's account, the world is determinate, and those conceptual contents simply are what they are in things" (Pinkard 2020, 15). In contrast, Pinkard's main point is that Hegel discusses the distinctive feature of humans that is thought. Humans do not primarily deal with the problem of a mind-external reality but rather explore and reveal themselves. Once they "extend [their] thought all the way to science" (Pinkard 2020, 16), Spirit becomes a "progressive Entäußerung—or emptying, pouring out—of our subjectivity in history" (Pinkard 2020,17). During this progressive externalization, we approach “modern Sittlichkeit”, which means that we operate on the base of making ourselves conscious. Nevertheless, even this ‘modern Sittlichkeit’ is only “the historical product of a certain Western tradition” (Pinkard 2020, 17). It thus remains unfinished and needs to be mediated in regard to the future history of humans. Overall, it means that Hegel does not argue for an objectivity that relies on a mind-external reality.

I agree with Pinkard’s criticism of Brandom’s account as ahistorical and Pinkard’s alternative notion of humans as historically self-approaching consciousnesses. My aim is to develop this notion with regard to Hegel’s *Phenomenology* as a relativist, empirical realism. My view on Hegel may be controversial for some philosophers. For others, it may knock on open doors. In both cases, however, it contrasts with Brandom’s account, thus bringing problematic aspects of his semantic, conceptual realism to the fore. In my view, Hegel ultimately suggests an alternative model of how to do philosophy without the standard of a mind-external reality, which I regard as the more viable philosophical option.

In the following chapters, I will discuss Brandom’s realism with regard to Habermas’ perspective. Habermas criticizes Brandom on similar grounds and demands a historical view of the problem. Interestingly, however, Habermas does not perceive Hegel as the historical thinker that I have in mind.

CHAPTER 3

Transition—Habermas' Contextualism, Brandom's Realism and Hegel's Historicity

Habermas' Discourse Theory holds relevance when addressing the divide between Continental and analytic philosophy, as both Habermas and Brandom strive to bridge the gap between these two approaches through a pragmatic theoretical framework. In the context of Höffe's interpretation, for example, Brandom is recognized as an adherent of this tradition of discourse theory:

> One of the most advanced contemporary theories of truth, the discourse theory developed by Robert Brandom in the spirit of American pragmatism, sets aside the (allegedly) descriptive question about the nature of truth and simply addresses the performative question about how the members of a linguistic community come to treat something as true.
>
> HÖFFE 2010, 186[1]

Höffe, as so many others, is excited about Brandom's achievements. Nevertheless, the characterization of Brandom's approach as discourse theory may be imprecise. The link between Brandom's inferentialism and Habermas' discourse theory is the "pragmatic integration of reference-semantics" (Braun 1996).[2] However, we must recognize that both aim for very different solutions to the problem of empty coherentism, the problem of how to secure truth without immediate access to the mind-independent world. The two solutions depend on the double-structure of utterances, namely the propositional part and the performative part. Either we stress a stronger attachment to mind-independent reality and thus focus on the propositional part, or we choose the path of our normative involvement in social and historical relationships and thus focus on the pragmatic conception of language. Though Brandom wants to walk both ways, he ultimately has a strong attachment to conceptual realism. In his book *Truth and Justification* [*Wahrheit und Rechtfertigung* (1998)], Habermas shows that Brandom attempts to combine these two possibly

1 Höffe, Ottfried. *Kant's Critique of Pure Reason: The Foundation of Modern Philosophy.* Tübingen: Springer, 2010.

2 Braun, Edmund. *Der Paradigmenwechsel in der Sprachphilosophie.* Darmstadt: Wissenschaftliche Buchgesellschaft Darmstadt, 1996.

 DOI:10.1163/9789004714809_004

incompatible approaches. He criticizes his attempt to solve the problem of an empty coherentism with his inferential, yet propositional semantics, which presupposes access to a mind-independent world.

For Brandom, as discussed before, the proposition is the downtown of language (Brandom 2010, 301). Habermas, in the course of correcting his own discourse-theoretical approach, rejects this position and chooses the opposite path: he develops a pragmatic, historical conception of truth, where the concept of a mind-external world is only the pragmatic result of discourses. The idea of a shared world remains therefore discourse-internal.

For Habermas, Brandom's conceptual realism cannot solve the problem of truth. Truth, as Habermas points out, is intrinsic to our pragmatic discourses since discourses occur in social communities of individuals who are conflicted about the status of their world when their practice fails. While Habermas argues that the concept 'world' remains discourse-dependent, Brandom's work proposes a formal way of communication that is centered on theoretical observers who indirectly address the status of *the* mind-independent world through their discourses. Brandom's approach, the idea that correct communication is grounded on a propositional structure, subordinates the normative realm of freedom to the self-movement of concepts as our true reality. Habermas claims that such, in his view, 'Hegelian' 'objective Idealism' is not feasible. For Habermas, there is no theoretical solution to the problem of the so-called efforts of the concept [Anstrengung des Begriffs]. Instead, truth remains the question of a historical practical discourse itself.

While I agree with Habermas' criticism, it is unfortunate that he also dismisses the possibility of a Hegelian continuation of the debate. Habermas perceives Hegel as an ahistorical thinker who envisions a predetermined unfolding of history based on reason. In this chapter, I aim to examine Habermas' critique of Brandom and its connection to Hegel.

It is worth noting that the debate between Habermas and Brandom took place two decades ago. Since then, we have witnessed an increased integration of Continental thinkers into the analytic program. This development suggests that we may have discovered a bridge between two formerly opposed intellectual traditions. However, the debate between Brandom and Habermas reveals that incorporating Continental thinkers into the analytical framework remains a challenge. Continental thinkers are often viewed as contributors to specific topics and themes to address the issues within the analytic tradition. For instance, Brandom's focus was on the question of metaphysical realism. Unfortunately, both Brandom and Habermas missed an opportunity for a genuine re-engagement with Hegel as a historical thinker.

1 Differences between Habermas' and Brandom's Account

Before I go into the details of Habermas' criticism, I want to give a brief characterization of what I take to be its main features. Although Habermas follows the insights of the linguistic turn, Habermas' theory remains at its core Continental. It is a Marxist theory, however, without Marxism (see Rockmore 1989, XI).[3] It means that, on the one hand, Habermas follows the Marxian idea that social practice interferes with our epistemological access to objects. Either it distorts the relationships of subjects to objects completely or, at least, partially. It thus interferes with our cognitive ability to construct a fair society. Unlike other Marxist programs, which usually demand revolution, Habermas believes that the increased critical discourse of real human beings will advance our societies towards Marxist ideals. Habermas rejects Marxism and its ideas of revolution. To him, it is still based on the subject's teleology from which we conclude an unjustifiable predestined history of humanity that demands revolution. It is still attached to Hegel's ahistorical philosophy that commits to the paradigm of a non-empirical self-consciousness with a very disputable teleology. For Habermas, a clear picture of the future that would be necessary for revolutions is impossible to obtain at this historical moment. Habermas therefore commits to a conservative ideal of improving current discourse structures and believes in a reformation of society.

Because of its critical positions regarding consciousness-paradigms and teleologies, Habermas' discourse theory remains close to the neo-pragmatic tradition. The linguistic turn seemingly offers a philosophy that can circumvent the problem of an unreliable, distorted consciousness and fulfills the demands of a post-metaphysical discourse. With his book *Theory of Communicative Action* [*Theorie des Kommunikativen Handelns* (1981)], Habermas therefore initiates a linguistic and pragmatic turn in Critical Theory's approach to societies. This turn seems to lead us away from metaphysical terms of world, soul, God, or truth, objectivity, reality and reference, and validity and rationality.[4] According to Habermas, a pragmatic, post-metaphysical theory delivers a "normative concept of linguistic cooperation," explains the "discursively validifiable claims," demonstrates what "formal pragmatic suppositions of the world" mean, and lays out "the conditions of rational acceptability" of the former term

3 Rockmore, Tom. *Habermas on Historical Materialism.* Bloomington: Indiana University Press, 1989, XI; Habermas, Jürgen. *Zur Rekonstruktion des Historischen Materialismus.* Frankfurt: Suhrkamp. 1976.

4 Habermas lists the same terms. I added the metaphysical terms *world, soul* and *God* (see Habermas, Jürgen. *Wahrheit und Rechtfertigung.* Frankfurt: Suhrkamp, 1999, 7).

(see Habermas 1999, 9). The cost of such a post-metaphysical program concerns the major difficulty of discussing a "world independent of our description" (Habermas 1999, 9). Since an "immediate approach to naked reality" is impossible, the question of how we ground our discourses emerges (Habermas 1999, 9).

We face the problem of so-called empty coherentism: if we have no access to reality, how do we justify our discourses? Habermas is critical of the solutions that the linguistic tradition has offered. According to him, the tradition of analytic philosophy has treated the propositional claim as the paradigmatic case of meaning and truth. Habermas notes that Frege and Russell, the semantic theories from Quine to Davidson, and from Sellars to Brandom pursue the mental theories of consciousness with analytical means (see Habermas 1999, 9). In many of these paradigms, the contextualist question of what is the "internal relation between successful communication and the representation of facts" is answered as if we have access to a mind-external reality (Habermas 1999, 11). With respect to this question, Habermas discusses Brandom's book *Making It Explicit.*

Like all great thinkers, Habermas sharpens his own approach during the study of other thinkers. His assessment of the Brandomian philosophy leads to some significant changes in his own approach. First, Habermas praises Brandom's *Making It Explicit* as a landmark [Meilenstein] (see Habermas 1999, 138). He compares Brandom's impact on philosophy with Rawls' *Theory of Justice.* He also notes that Brandom "painstakingly works out an innovative connection of formal pragmatics with inferential semantics" (Habermas 2003, 131).[5] The crossover of formal pragmatics and inferential semantics is the central question of the "pragmatic turn" (see also Habermas 1999, 170), the major reason why Habermas is so much in favor of Brandom. Yet, despite these appraisals, Habermas criticizes Brandom for widely ignoring the second-person perspective. According to Habermas, he understands a sentence only with regard to the theoretical conditions of an observer: we know what would be the case if the sentence were true (Habermas 1999, 11). This focus on the material consequences of utterances represents a shift away from the pragmatic conditions of utterances toward the perspective of the theoretical observer. Since, however, the access to the conditions of truth is not immediate, Habermas remarks that the central point is not the "solipsistic approachable conditions of truth," but

5 "geduldig-präzise Ausarbeitung einer innovativen Verschränkung von formaler Pragmatik und inferentieller Semantik" (Habermas 1999, 138). Translation by Barbara Fultner (Habermas, Jürgen. *Truth and Justification.* Translated by Barbara Fultner. Cambridge, Massachusetts: MIT Press, 2003).

the condition of what can be accepted as publicly rational (see Habermas 1999, 11). 'Rational' means here that our reasons can be shared publicly in a community. Nobody can be rational just by himself. The mastery of a concept means therefore to know how to use an expression "in order to cooperate with others on something" (Habermas 1999, 11). It indicates that there must be an "internal relation" between successful communication and the representational function of language. This internal relation, as Habermas further argues, is neglected in the framework of Brandom since he pursues a theoretical, methodological individualism (see Habermas 1999, 11). It means that Brandom's discourse participants do not cooperate, and do not consider the other party as a discourse participant, as the second person, but they only observe tactically with respect to semantic vocabulary and its material consequences. The formalist account of Brandom is therefore outside of the practical discourse and does not capture how a moral community achieves the shared perspective of 'us'.

Brandom's book *A Spirit of Trust* responds indirectly to Habermas' criticism. While *Making It Explicit* discusses concepts as a part of an unhistorical language game, *A Spirit of Trust* discusses the involvement of actors with regard to their societal position and normative involvement in a historically developing society. Even though this answers some questions about the historicity of human beings within their community, Brandom still relies on the transcendental language-game. So, in a sense, Habermas criticism that Brandom focuses on a tactical observer has to be renegotiated. I think, however, that the more important aspect is Habermas' criticism of Brandom's mind-external realism. With his criticism in his book *Truth and Justification,* Habermas steps back from his *Theory of Communicative Action,* seeking to assess somewhat objective conditions. While Brandom places semantics and a mind-independent reality at the core of his theory, Habermas moves through his criticism toward pragmatism.

Habermas' criticisms are motivated by a major shift in his own theoretical program. Consider how he addresses the problem of reality, which he analyzes in *Truth and Justification*: "The reality facing our propositions is not 'naked', but is itself already permeated by language" (Habermas 2003, 36).[6] So Habermas denies that we have direct access to a mind-independent reality. Rather, reality is indirectly grasped in our language. The truth of a statement is then mediated only with regard to other sentences. Here, Habermas identifies the main problem of contextualism:

6 Habermas 1999, 48.

> It seems that the truth of one proposition can be warranted only by its coherence with other, already accepted propositions. Yet neither the assumption of epistemological realism, nor the power of learning processes to revise from within the context in which they arise, nor the universalist import of context-transcendent claims to truth can be reconciled with a thoroughgoing contextualism.
>
> HABERMAS 2003, 36[7]

According to such a frictionless coherentism, claims can only be justified by already accepted claims. The problem is that we might have a coherent system, but our system could still be false. Habermas' solution to this problem is found in his earlier work with the notion of an ideal discourse:

> The attempt to combine the language-transcendent understanding of reference with a language-immanent understanding of truth as ideal assertibility promised a way out of this dilemma. On this view, a statement is true if and only if, under the rigorous pragmatic presuppositions of rational discourse, it is able to withstand all efforts to invalidate it, that is, if and only if it can be justified in an ideal epistemic situation.
>
> HABERMAS 2003, 36[8]

Accordingly, we achieve truth through processes of communication guided by ideal assertibility. In Habermas' case, it means that a claim is true if it can be defended against all attempts to deny it under ideal circumstances. We accept something as pragmatically true if we can defend it against all reasonable objections during an actual discourse. In its practice, truth depends on the inclusion of all concerned members of the communication-community who have equal rights to communicate. Habermas specifies these rules: all concerned persons have to be included, there have to be equal rights for communication, there is no violence so that only the forceless force of the better argument prevails, and all participants are truthful (see Habermas 1999, 49). In other words, truth is achieved by securing the social conditions for truth to occur. However, the question is whether the securing of ideal speech circumstances will actually achieve truth, or whether it only constructs another context. The possibility of a confirmation bias that exists between the discourse participants cannot be excluded. Even under ideal conditions of the discourse, they might agree

7 Habermas 1999, 48.
8 Habermas 1999, 48.

on something that is simply false and they might have no means to correct their false beliefs by merely proceeding with the discourse. As a result, we see Habermas drastically changing his position.

Habermas attempts to repair his procedural concept of truth as justifiability. He acknowledges first that truth is not a question of success: "The procedural conception of truth as discursive redemption of truth claims is counterintuitive to the extent that truth is obviously no 'success concept'" (Habermas 2003, 37).[9] There is no conceptual relation between truth and rational justifiability (see Habermas 1999, 50). Justification does not guarantee truth, and better discourses do not necessarily lead to better insights into how the world functions. Habermas therefore changes his concept of truth:

> These objections have prompted me to revise the discursive conception of rational acceptability by relating it to a pragmatically conceived, non-epistemic concept of truth, but without thereby assimilating 'truth' to 'ideal assertibility'.
>
> HABERMAS 2003, 38[10]

The new pragmatic conception of truth, which Habermas offers, means that an actual living being having no doubts about truth 'in actu'. In a word, truth is part of practice. Epistemic truth, the moment when we discuss the truth of a statement, only appears in the context of an opponent challenging our everyday practices (see Habermas 1999, 52). The question of truth is thus derived from our normal course of action, but not the standard case of communication.

How does this pragmatic concept of truth work? According to Habermas, actors have pressure to act. This pressure leads them to processes of learning. Only if an acquired belief can be justified by a process of acquisition, they can defend this as true knowledge for acting (see Habermas 1999, 55). Truth occurs when we reflect on our actions. Habermas writes: "Actors capable of dealing with the world feed on their certainties of action. But for subjects who reflectively ascertain their knowledge in the context of discourses, a proposition's being true and its fallibility are two sides of the same coin" (Habermas 2003, 41–42).[11] In other words, actors presuppose a mind-independent world only after the experience of practical failure. So, it is not about epistemic truth, but it is the pragmatist question of how truth functions in our discourses. With regard to Habermas' solution, Levine remarks that "objectivity is rendered in

9 Habermas 1999, 50.
10 Habermas 1999, 51.
11 Habermas 1999, 55.

the first instance not by our linguistically representing the world but through our acting upon it and within" (Levine 2011, 247).[12] Confronted with failure, we correct our practice since the world does not collaborate (see Levine 2011, 247). Only after failure, we will represent the world as "a totality of objects" in order to discuss a shared world (see Levine 2011, 247). We still do not discuss facts but objects resulting from a discourse. Truth is therefore a result of practice. The positive aspect of Habermas' theory is that though truth occurs in discourses and its status is diminished, it is not rejected overall.

With his practical account to truth, Habermas proposes a model of experience, according to which practice is primary. Theories will be built on the basis of our processes of experiences, while truth is a part of our practical way of living (see Levine 2010).[13] Altogether the question emerges: Does Habermas give up on his former concept of Discourse Theory and does he come closer to a Hegelian conception of truth that focuses on the moment of revising truths through experience in a social community? Though Habermas comes closer to the relativist Hegelian account of truth that I defend (especially in his criticism of Brandom), he denies that it is a return to Hegel. He does so for two reasons, first because of his own rejection of metaphysics. Second, he rejects what I argue is a misconception of Hegel. Independent of his rather pragmatist conception of truth, the progress of analytic philosophy is for Habermas a linguistic movement to replace the paradigm of consciousness and to develop a non-metaphysical philosophy that resembles the supposedly non-metaphysical, empirical operations of science. Habermas therefore persistently holds on to his philosophy that is in line with his early rejection of consciousness-philosophy:

> The appropriation of hermeneutics and linguistic analysis convinced me then that critical social theory had to break free from the conceptual apparatus of the philosophy of consciousness flowing from Kant and Hegel.
>
> HABERMAS 2015, 7[14]

12 Levine, Steven. "Truth and Moral Validity: On Habermas' Domesticated Pragmatism." *Constellations* 18 (2), 2011: 244–259. https://doi.org/10.1111/j.1467-8675.2011.00636.x.

13 Levine, Steven. "Habermas, Kantian Pragmatism, and Truth." *Philosophy and Social Criticism* 36 (6), 2010: 677–695. http://journals.sagepub.com/doi/abs/10.1177/0191453710366216.

14 Habermas, Jürgen. *On the Logic of the Social Sciences.* Translated by Shierry Weber Nicholson and Jerry Stark. Cambridge: Polity Press, 2015.

Habermas' counter-approach to these theories of consciousness is supposed to be sociological; a method that he believes does not rely on the internal representations of objects. Instead, sociology only describes human interactions on the societal level. From his description of his theory, it becomes clear that Habermas follows the historical trend of the early analytical movement to reject Hegel as metaphysical.

We can sum up Habermas' theory as a combination of the continental Marxist rejection of Hegel, in the form of a quasi-Marxism, and the analytical reaction to Hegel on account of his anti-metaphysics, i.e., analytic pragmatics. Habermas further rejects a transition to the Brandomian-Hegelian framework, because Brandom reduces our cooperative efforts to demonstrate the truth of practice in discourses to a task of analyzing our theoretical commitment to material consequences of utterances. According to Habermas, it is the same mistake Hegel makes, when he unburdens individuals from their individual responsibilities and replaces their responsibility through a concept of a predestinate world spirit.

Habermas' criticism of Brandom's reliance on semanticism is concise, Habermas' reconstruction of Hegel, however, is based on the assumption that Hegel has a closed system, while all development is guided by an ahistorical, absolute spirit. My goal is therefore to offer an interpretation of Hegel that respects Habermas' criticism of Brandom but proceeds with Hegel's account of a system that is open towards history. Before presenting this interpretation of Hegel, I will reconstruct Habermas' criticism of Brandom and Hegel in detail.

1.1 *Pragmatics and Semantics—Two Incompatible Approaches in Brandom?*

The question for the contextualist is how to secure knowledge within a pragmatic language without immediate access to a mind-independent reality. Habermas reconstructs Brandom within the confines of addressing this question. Brandom wants to solve the problem from two sides. On the one hand he wants to explain content in terms of pragmatics. On the other hand, he wants to explain the semantic structure of our claims that confers content. Giovagnoli identifies here the "ambiguity […] between pragmatics and semantics;" it is a contradiction "that social practices confer conceptual content to states and expressions and that material rules of inferences confer, at the same time, that content" (Giovagnoli 2001, 54).[15] Habermas finds sympathy with this line of

15 Giovagnoli, Raffaela. 2001. "On Normative Pragmatics: A Comparison between Brandom and Habermas." *Teorema* 20 (3): 51–68. Accessed November 26, 2017. https://www.jstor.org/stable/43046449.

critique, and focuses on the incompatibility of both approaches: pragmatics and semantics. In the following paragraphs, I will discuss Habermas' reconstruction of Brandom's approach and carve out his criticism that Brandom pursues two incompatible approaches.

According to Habermas, Brandom's approach to communication is based on the Wittgensteinian model that there is a "practically pre-predicative knowledge" (Habermas 1999, 140). It means that we first act as members of a social community and then we identify our actions by means of predication. In other words, pre-predicative knowledge precedes the propositional, logical, and expressive vocabulary. Moreover, the social practices resulting from this pre-predicative knowledge can even precede the intentions of individual speakers.

Explicit and implicit structures are both necessary components of language, since without the competence of making our practices explicit, there would be no practice that could be called rational. Although this linguistic competence is mostly unspoken and implicit, at the same time, we are always capable of making our linguistic competence explicit. Hence, rational beings are logical, expressive beings for Brandom (see Habermas 1999, 140). By virtue of Brandom's use of expressive vocabulary, "epistemic authority" is then transferred from the private subject to the public practice of the community (see Habermas 1999, 140). This authority does not rest on internal representation of contents but is about the community-guided understanding (Habermas 1999, 141). In other words, we communicate during practice, for example, when we build houses, prepare food, or create a common social space. These practices are not related to the question of how we represent the world. Thereby, however, communicators find themselves always in a "net of interactive relations", in which they exchange reasons (Habermas 1999, 141). For Brandom, the exchanges within this net are the practice of giving and asking for reasons.

The practice of giving and asking for reasons entails that we are not merely listeners, but that we adopt the attitudes of the speaker, in order to understand a sentence (See Habermas 1999, 142). Since reasons cannot be understood without knowing their "weight" [Gewicht], the listener also has to actively judge the entitlement of the speaker (see Habermas 1999, 142). From here, we acknowledge the credibility of the speaker, or in Brandom's words, we keep score (see Habermas 1999, 144). We do so by tracing the material consequences of what the speaker says. Since the statements must be structured inferentially, we can assign commitments and entitlements to speakers. We understand the conditions of an utterance by judging the material consequences that statements, as parts of inferences, entail. If we agree with the claims of a speaker and see it in accordance with inferences, we keep a positive score of the speaker's credibility. Brandom therefore develops a complementary semantics, which Habermas signifies as his greatest achievement, but which will also turn out to be a major point of criticism (see Habermas 1999, 145).

The strong focus on practice is supposed to avoid representationalism and pays attention to the self-binding rules that speakers oblige themselves to whenever they communicate with each other. We see that with regard to this rather relativistic, cooperative perspective on language, that "being a reason is to be understood in the first instance in terms of what it is for a community to treat something in practice as such a reason [...] as reasons for claims" (Brandom 1998, 253, see also Habermas 1999, 140). Social practice determines the status of reasons. Nevertheless, as this citation also shows, it is not only social practice that determines what counts as reason, but that reasons are also linked to propositional claims. Claiming, as Brandom says elsewhere, is the downtown of language (Brandom 2010, 301). Speakers therefore use language to make the underlying propositional structure of their claims explicit.

The first critical observation I think is important to highlight is Habermas' concern for external rules. With respect to the formal language model that Brandom develops, the theoretical observation and assignment of credibility to the second speaker occurs much as it does in, say, a game like chess. Here, opposing players agree on external rules. While playing, they observe each other strategically without really interacting with each other as responsible persons. Unlike chess, a more realistic approach to communication suggests that speakers follow a variety of rules that can change. In this case, rules are altered with regard to the contingent contents of their historical community. Sometimes new practices are adopted, and sometimes the practices are altered. The way the game of language is played depends on the historical situation and its practical demands. Moreover, players do not just play against each other but also cooperate.

However, before we criticize Brandom's formalism in this way, we have to keep in mind that Brandom develops only a model for describing the communicational situation in *Making It Explicit*. In *A Spirit of Trust,* Brandom broadens his view and discusses how the practice of forgiving creates a social community of trust. However, as the sole criticism of the model character of Brandom's approach would be too simple, Habermas focuses less on the formalist perspective that Brandom develops in *Making It Explicit*. He rather criticizes that only ascribing material content to statements by a normative pragmatic, which is Brandom's central idea, is not enough to solve the problem of empty coherentism, the problem that also concerns Habermas' own theory. Thus, he asks critically:

> If the practice of mutually attributing and assessing truth claims cannot already be guaranteed through the semantic establishing of materially valid inferences, then of what kind are the constraints on truth? Something or other has to corroborate the correctness of the application of concepts—'the assessment of truth'.
>
> HABERMAS 2003, 138[16]

Just to keep score of linguistic actions does not guarantee that our theories are correct. We need to demonstrate that our actions are anchored in reality. To solve this problem, Brandom assumes that concepts are understood within the boundaries of normative pragmatics. He thus transforms the question of truth into the question of norms. If our norms are not anchored in reality, then our language games would be meaningless. Though we cannot approach reality directly, we pragmatically acknowledge the necessity of norms being anchored in reality. Habermas is skeptical. Norms are not naturally part of all humans interacting with each other since there are no norms without members of a community who acknowledge them (see Habermas 1999, 147). So the question emerges: can we justify norms if the instance of 'the law-giver' is inside of the laws (see Habermas 1999, 148)? The autonomous creation of laws cannot operate without being circular: "Norms have to be established "rationally" in accordance with norms of reason, and this process therefore cannot itself provide the model for an explanation of normativity" (Habermas 2003, 140).[17] Habermas emphasizes that the law-giver is already within a historical society that gives orientation for such institution of norms. Of course, in his book *A Spirit of Trust,* Brandom discusses the difficulties of relative laws. He explicitly acknowledges that the creation of laws on the basis of the modern, subjective acknowledgment leaves no room for error and would run into the dilemma of the master: losing touch with reality as he just claims authority without responsibility. So the creation of norms depends for Brandom on the intertwinement of authority and responsibility.

In *Making It Explicit,* Brandom does not discuss the law-giver as embedded in a historically developed society. Instead, he claims that we come to the insight that our language games are normatively grounded in a transcendental language that is based on assertions (see Habermas 1999, 154). The assertion is then the ground for any kind of rational activity, and is free from cultural norms or historical incompleteness. But how can he possibly prove this? Habermas

16 Habermas 1999, 146.

17 Habermas 1999, 148.

writes on this contradiction: "[the] 'realist' objection, which Brandom seems to raise against himself, is hardly consistent with a 'phenomenalist' stance" (Habermas 2003, 138).[18] This means that a pragmatic construction of language with respect to norms and a semantic, formal realism that presupposes contact with reality are probably incompatible. This problem will also persist with regard to Brandom's approach in *A Spirit of Trust*. The presupposition of a mind-external reality is a central part of what he understands as conceptual realism. Before Habermas, however, draws conclusions on the impossibility of this endeavor, he investigates Brandom's realist, pragmatic solution in further detail.

1.2 *Anaphora as an Anchor in Reality*

According to Habermas, Brandom follows a transcendental, linguistic deduction of the "double-structure" of utterances based on simple, predicative assertions (see Habermas 1999, 154). Now, Habermas wonders whether we can ground pragmatics in semantics, and investigates Brandom's conceptual realism with respect to this question. For this, he analyzes Brandom's idea of anaphora which is supposed to show how statements are anchored in a shared reality.

For Brandom, assertions are the smallest units of meaning. Yet, they do not simply reference a mind-independent object. They rather serve as conclusions or premises in inferences and can indicate, in particular cases, the systematic place of deictic events, i.e., events that we point to, in our semantic net by employing anaphora. Brandom follows the intuition that recognizing something once is impossible. In this case, we could not recognize the objects under different circumstances again. He therefore assumes that the linguistic capacity to build anaphoric chains allows for a recursive relation to former moments of perception, i.e., deictic events (see Habermas 1999, 155). This implies further that without anaphora, there is no deixis since pointing out something would remain an unrepeatable event and the deictic action would be then useless. By anaphoric reference, the original deictic event can be located and placed in our linguistic system. Habermas writes on this aspect: "Only the intralinguistic reference to antecedent parts of the sentence makes it possible to refer to objects that, going beyond individual deictic acts, we have to be able to hold onto as reidentifiable objects" (Habermas 2003, 146).[19] This anaphoric relation is supposed to secure objectivity. Objectivity is for Brandom the difference between

18 Habermas 1999, 146.

19 Habermas 1999, 156.

what people believe to know and what they really know (see Habermas 1999, 157). In particular, the interpreter ascribes a commitment *p* to the speaker but distinguishes the truth of *p* from the speech-act of claiming *p*. In this ongoing ascription, participants create an interpersonal map of reality by finding warranting reasons that are supposed to make claims reasonable. The last warranting factors of our communication are deictic events that can be traced back through anaphoric chains. Anaphoras therefore indicate original deictic events.

Though cognition is not limited to these operations, they are supposed to get the conceptual system off the ground. Habermas, however, does not think that this is the solution to the problem of contextualism, the possibility that all communication-participants could be wrong. This is because there is no guarantee that our words are necessarily anchored in what is relevant. Habermas asks: "Is this kind of anchoring in itself sufficient, however, to satisfy the realist intuition of an independent world that can challenge even our best description" (Habermas 2003, 150)?[20] He particularly remarks that such empirical anchors and the navigation around points of reference cannot explain why there are new norms developing. So, for example, even if parents could teach their child to reliably respond to the experience of blue with the word 'blue', then it would still remain unclear how children could radically change their vocabulary about such perceptions (see Habermas 1999, 161). Habermas thus indicates that learning includes more than simply locating our references in a semantic framework. Though "perceptions [...] mark a point of intersection between language and the world," it does not specify the "veto power that an objective world can exert" (Habermas 2003, 150). Can we, for example, revise our concepts of 'blue' with our current "empirical standards" and acknowledge that 'blue' is now something else? Habermas does not see that experience alone has such "a far-reaching revisionary power" (Habermas 2003, 150).

In my view, Habermas supports my arguments that it is not about an external world as the ultimate verifier, but that it is instead about how we experience and learn within a social and historical community. Experience involves a dialectic of thought and perception in relation to the question of history. It is not only about trained anaphoric chains of reference, but also reasons why we historically and socially draw our attention to certain experiential events. Brandom's conceptual framework has no paradigm shifts. It is grounded in the premise of an external reality and one system of communication. Even Brandom's amendments in *A Spirit of Trust* do not sufficiently address this

20 Habermas 1999, 159.

problem. Although it is true that Brandom considers the historical dimensions of a developing community of reason that employs changeable concepts, he still constructs an ahistorical framework for these concepts. This framework is a community of trust that forgives necessary mistakes occurring through the *one* process of negotiating mind-external reality. Their process relies on Brandom's modal and conceptual realism which means that they somehow attempt to clarify the relationship between their phenomenal experience and the noumenal realm of things in themselves.

For Habermas, Brandom's proposed anaphoric solution of *Making It Explicit* is insufficient. It cannot exhaustively explain how perceptions lead to a revision of semantic concepts (see Habermas 1999, 162). Simply presupposing that our conceptual framework will be falsified or even verified by nature is a strong presupposition and seems to lean into the direction of naturalism. However, as already explained above, Brandom pursues a non-naturalist strategy (see Habermas 1999, 163). For Habermas, the problem of naturalism is its translation of everything into terms of causality and neurological states, which establishes an unbridgeable gap between the original phenomena and its more, and more detailed, but more and more unintelligible interpretation of the original phenomena (see Habermas 1999, 162). So, for example, a table could be analyzed into its component parts down to its almost atomic nothings, while at the same time the phenomenological character of the table is not expressed by these physical concepts. On the contrary, Habermas asks how the perceived events should be described by virtue of a nominalist language, a language in which people share their practical perspectives and construct a shared world (see Habermas 1999, 163). For Habermas, the relation to the world is therefore insufficiently discussed in Brandom (see Habermas 1999, 163). This criticism applies to *Making It Explicit* as it also applies to *A Spirit of Trust.*

In regards to the question of the revision of concepts, Habermas also discusses Brandom's example of the litmus-paper-test. Brandom presupposes that something tasting sour but not turning litmus blue must account for our practical failure. For Habermas, the litmus-paper-test only explains how we revise concepts that are already known: "Here, perception is generated through an experimental action. Such an experiment, however, merely exploits the internal connection between perception and action that already exists in everyday practices and that makes possible ordinary 'learning'" (Habermas 2003, 154). So, for example, we could imagine societies that would not acknowledge the litmus paper turning blue as an indication of our practical failure because they have simply no practices in which this test should be significant. Habermas therefore proceeds with his critique of Brandom's presupposition of a language-external world:

> Only if agents distance themselves from their practical coping with the world and enter into rational discourse, objectifying the situation that was originally 'ready-to-hand' in order to reach understanding with one another about something in the world, can such a perception become a discursively mobilized 'reason'. It then enters as criticism into the conceptual economy and the semantic inferential resources attached to existing views, setting in motion revisions, if necessary.
>
> HABERMAS 2003, 154–155[21]

For Habermas, an objectifying discourse turns experiences into experiences of failure. For *A Spirit of Trust,* Brandom presupposes the objective discourse as the way of how we negotiate in general. It is not a derivative from our *Lebenswelt*, but the formal ground for our interaction. Without considering the dimensions of an objectifying discourse, Brandom moves linearly from perception to action (see Habermas 1999, 165), as if there is a predetermined path, and fails therefore to develop a complete theory of learning that can also account for paradigm shifts in the formal semantic model (see Habermas 1999, 165). Habermas judges for that reason: "He does not conceive of the world we encounter in any way nominalistically but rather [...]'realistically'" (Habermas 2003, 155–156).[22] In other words, Habermas criticizes Brandom's solution as a conceptual realism. It is the idea that the concepts of our contingent language are developed only with regard to the mind-independent structure of the world that we 'simply' perceive. The phenomenon of interpretation would then be something by which real, mind-external things are distorted. Habermas cites Brandom with regard to his conceptual realism:

> The conception of concepts as inferentially articulated permits a picture of thought and of the world that thought is about as *equally*, and in the favored cases *identically*, conceptually articulated.
>
> BRANDOM 1998, 622

Brandom assumes that the world and thought have an identical structure. Experiences of the world only have a passive role in shaping our conceptual understanding of the world (see Habermas 1999, 166). The human mind does not answer creatively to disappointing stimuli but receives a passively imprinted, and yet distorted picture of the world (see Habermas 1999, 166).

21 Habermas 1999, 164.

22 Habermas 1999, 166.

With the hermeneutic turn, Habermas introduces a contrary, Continental concept of language that deals with the disadvantages of nominalism, namely that the world is only a practically presupposed concept. Hermeneutics, according to Habermas, integrates the world as an inner-worldly challenge. Accordingly, humans interpret historical occurrences within their language (Habermas 1999, 167–168), in contrast to Brandom who turns concepts into objective things that are related to the way things really are. The creative functions of language remain for Brandom outside the core of language and are thus unintelligible. The assertion expresses rationality, while other symbolic forms of arts must be entertainment and occupation of an overly active mind. Habermas cites Brandom with regard to this objectifying conceptual realism:

> Concepts conceived as inferential roles of expressions do not serve as epistemological intermediaries, standing between us and what is conceptualized by them. This is not because there is no causal order consisting of particulars, interaction with which supplies the material for thought. It is rather because all of these elements are themselves conceived as thoroughly conceptual, not as contrasting with the conceptual.
>
> BRANDOM 1998, 622

The world is, in this sense, still the completeness of facts (Habermas 1999, 169). According to Brandom, we only have to find the right way to mediate this world. The fundamental structure is our reliance on assertions.

Habermas' major criticism is therefore that Brandom's model of anaphoric anchors does not consider a possible revision of the semantic rules (Habermas 1999, 160). Brandom thus does not only use a formal model but constrains himself to this formal model. The concept of a world that is intervening in our actions is for Brandom considered only with respect to an instrumental language. Yet, the concept of an independent world of objects is still produced within language. As a consequence, the concept 'world' is not mind-independent. If its concept changes our concept of communication might change as well.

1.3 *Brandom's Leveling of Facts and Norms*

Firstly, Habermas does not accept Brandom's approach to mind-external reality as compatible with his 'phenomenalism'. Secondly, he criticizes Brandom's model of communication. In his third point of criticism, Habermas denies that Brandom successfully integrates the second-person perspective. He argues that Brandom introduces a third-person, the perspective of a neutral observer that, however, disqualifies the moral feelings of the involved actors. By this,

Brandom dismisses the importance of people achieving truths in a discourse by virtue of their moral feelings. Habermas concludes therefore that the main premise of *conceptual realism* unburdens human society from its constructive and subjective tasks. It thereby levels facts and norms:

> to a certain extent the supposition of a world that is entirely conceptually structured relieves the finite and fallible human mind of the burden of the constructive endeavor to develop its own concepts in terms of which it can interpret what happens in the world.
>
> HABERMAS 2003, 161[23]

As Habermas explicates further, for Brandom, the objective content of concepts means to make them explicit. It presupposes that the world in-itself is rational (conceptual) and so it also limits our normative discourses. Our individual feelings as participants are then subordinated to the objective explications of our assertions.[24] For Habermas, Brandom therefore levels the gap between subjects who operate normatively and objects that relate causally to each other. Brandom writes, indeed: "fact-stating talk is explained in normative terms, and normative facts emerge as one kind of fact among others" (Brandom 1998, 625–626). For Brandom, language is just another fact among other facts. Does this sacrifice moral autonomy for conceptual holism? For Habermas, the 'effort of the concept' that subjects carry cannot be simply replaced by the 'self-movement of the concept' that develops according to a factually motivated morality. Habermas emphasizes this again:

> The 'effort of the concept', which otherwise would be a matter of cooperative learning by way of the constructive interpretations of a communication community, is replaced by a 'movement of the concept', which proceeds through discourse and experience but over the heads of most

23 Habermas 1999, 172.

24 Habermas constructs here a first relation to Hegel. Hegel who often is held to uncritically assume that the world is rational [vernünftig], writes, in fact: "To him who looks upon the world rationally, the world in its turn, presents a rational aspect" (Hegel 1861, 11). It must be further investigated in how far the world is rational because we take it to be rational. Thus, the question is more complex and bound to our subjective perspective. The main task for Hegel is to demonstrate how something that is subjective can still yield objective cognition. For this reason, Hegel's objective Idealism is not a conceptual realism (Hegel, G.W.F. *Lectures on the Philosophy of History*. Translated by J. Sibree. London: Henry G. Bohn, 1861).

> of the participants. Objective idealism divests them of the epistemic authority (and also of the moral autonomy).
>
> HABERMAS 2003, 161[25]

Central to Habermas' criticism is that Brandom transfers the burden of moral evaluation from the involved participants to the neutral observer. With regard to this aspect, Habermas discusses Brandom's example of a trial. Accordingly, the judge and the jury observe and evaluate the arguments of the prosecutor and the defender. It is of lesser interest how the participants judge their positions and how they respond (see Habermas 1999, 174, and Brandom 1998, 505). This is significant since a statement that does not take the real answer of the participants into account underestimates the grammatical function of the second person. Brandom's formal account investigates, according to Habermas, only the formal, doxastic commitments. Thus, Habermas states: "A research strategy that confuses the first level of communication with the second ignores, in overlooking this important distinction, the grammatical role of the second person" (Habermas 2003, 163).[26] Consequently, he calls this dismissal of the alter-ego "theoreticism" (Habermas 1999, 174). It means that communication is the result of an epistemic relation between the assertions of the speaker and the ascription of the interpreter about what has been said (see Habermas 1999, 174). On the contrary, Habermas points out that the utterance of the speaker is, at the same time, an imperative to coordinate different opinions, different backgrounds, and different interpretations of what it means to live in a shared world, especially since 'world' is an intersubjective product of our discourses. It is, moreover, not only a theoretical battle of systems or a choice of forms of expression. It is rather a necessity of coordination that brings us into the play of language games (see Habermas 1999, 175). It is neither a self-sufficient [selbstgenügsames] play, nor a question of whose standpoint will prevail (see Habermas 1999, 175). Habermas sees the most important evidence for this coordination in the fact that we expect answers, if we claim something (see Habermas 1999, 174–175). For this reason, Habermas suggests understanding our speech acts with regard to the goal of coordination: "We understand a speech act when we know the conditions and consequences of the rationally motivated agreement that a speaker could attain with this speech act" (Habermas 2003, 165).[27] Here, the acknowledgment of somebody's utterance does not concern the material consequences of the concept that he

25 Habermas 1999, 172.
26 Habermas 1999, 175.
27 Habermas 1999, 176.

uses, which expresses Brandom's inferentialism. Instead, it brings each communication back to the question of real human persons whose intentions are to understand the other and to coordinate their views. While the score-keeping model of Brandom is only about strategic observation, it is about coordination between discourse participants in the model of Habermas. For Habermas, Brandom's model expresses a methodological individualism that fails to understand that participants work together, instead of working against each other (Habermas 1999, 177).

Brandom ties communication to the objective representation of facts in the world that only have to be traced in terms of their logical place in the semantic net. By this, we make the real status of the world explicit. This asserting seems to settle things, and makes communication seemingly easier, but it rests on the assumption that truths settle themselves (see Habermas 1999, 177). For the very same reason, Habermas rejects all forms of objective idealism and states finally: "A tacit return to objective idealism can scarcely be possible anymore" (Habermas 2003, 159).[28] In other words, Habermas also rejects Hegelianism.[29]

According to Habermas, the leveling of facts and norms leads to a moral realism that is grounded in logocentrism. On the contrary, Habermas points out that the social background allows for manifold speech acts that possibly cannot be subsumed under the idea of the assertion (see Habermas 1999, 178). Because of the question of objectivity, however, Brandom commits to conceptual realism, the idea that conceptual structures of the world are coining our practices of discourse (See Habermas 1999, 177–178). Brandom writes: "Concepts are rules, and concepts express natural necessity as well as moral necessity" (Brandom 1998, 624). The Brandomian interpretation that everything is normative and that the normative can only be justified by assertions intertwines the realm of freedom ["Reich der Freiheit"] with the causal necessities of the world (see Habermas 1999, 179). Though Brandom includes both perspectives, he favors the semantic side. Brandom's "conceptual monism" only discusses the theoretical responsibility but does not discuss the question of how we bind ourselves with regard to rules of assertion (Habermas 1999, 182). Habermas explains the difference: "Although norms of action can be described as facts from the observer's perspective, they can be justified only from the participant's perspective. This is evident from the very form of practical inferences, which refers to an 'I'" (Habermas 2003, 170).[30] For Habermas, the participant's

28 "Ein *stillschweigende* Rückkehr zum objektiven Idealismus dürfte kaum noch möglich sein" (Habermas 1999, 170).

29 I will clarify later that Hegelianism is not a conceptual realism.

30 Habermas 1999, 182.

perspective of speech acts evidently includes another dimension that possibly cannot be objectified. The self-reference of a speaker, the I, is an existential adoption of responsibility within a communication-community that can only be reflected upon after the fact. The moment of responsibility, however, is an existential burden that is not adequately grasped by responsibilities induced through factual justification.

Beyond this emphasis on the second-person, Habermas gives further reasons why it makes sense to distinguish a practical will from factual knowledge. According to Habermas, Brandom does not take into account the differentiation between the will's heteronomy and the will's autonomy. He distinguishes between subjective reasons and instrumental knowledge: "In the case of a purposive-rational choice of means, the rules adopted by the agent establish the connection between a subjective reason for acting and an instrumentally employed factual knowledge" (Habermas 2003, 171).[31] So, in the example of opening an umbrella, prudence is the result of an empirical rule that is subjective with regard to the contextual conditions of the surrounding world of facts. It leads to the rational insight that I should open my umbrellas when it rains. Facts determine actions. Yet, it is different with regard to questions of morality, as Habermas explains: "Whereas prudential and conventional reasons bind free choice only relative to given interests and existing social values, moral reasons claim to penetrate the will completely, that is, to determine it absolutely" (Habermas 2003, 171).[32] In other words, for moral actions we claim universal validity independently of our subjective interests, a validity that goes beyond something that is related to the world of contingent facts, but that specifies how something should be. The more we distance ourselves from the contingent facts that motivate our behavior, the more we act autonomously. For example, we all could be biologically determined to eat meat. Yet, even if this were a fact, it would not be sufficient to universalize the action of eating meat. One might argue that eating meat is immoral, even though it is biologically advantageous. Although facts and norms have similarities in their justification, Habermas denies therefore: "Whereas prudential and conventional reasons bind free choice only relative to given interests and existing social values, moral reasons claim to penetrate the will completely, that is, to determine it absolutely" (Habermas 2003, 171). According to Habermas, and as we see in the example of a meat-eating society, facts cannot deliver a sufficient reason for decisions (see Habermas 1999, 184). Brandom, on the contrary,

31 Habermas 1999, 185.
32 Habermas 1999, 184.

clearly states: "In this way the normative is picked out as a subregion of the factual" (Brandom 1998, 625). However, morality is about the question of how we can achieve the universal character of moral deeds, while universality is not only a question of facts.

Habermas' answer is that universalization must be related to fair procedures that have to be socially secured with regard to all potentially concerned ones [Betroffene]. Including the 'concerned ones' guarantees, at least, the inclusion of the participant's perspective. Habermas' solution has its own problems, since the guarantee of fair procedures does not necessarily lead to better societies. With regard to Brandom, however, he concludes convincingly:

> The deontological understanding of morality, which Brandom himself also favors, does not fit the conceptual-realist understanding of the moral vocabulary he proposes in order to anchor the objective content of our concepts [...] in the conceptual structures of the universe.
>
> HABERMAS 2003, 173[33]

Since the world of facts does not deliver a clear moral criterion for our universalization processes, we need to consider our autonomy and freedom and include them in a dialectical investigation. All in all, we can say that though Habermas praises Brandom, he ultimately rejects his account.

2 Habermas and Hegel

Habermas' criticism of Brandom's conceptual realism leads him to a specific perspective on Hegel. In fact, he equates Brandom's 'conceptual realism' with Hegel's 'objective idealism' (long before *A Spirit of Trust* was published). Habermas believes that Hegel defends an absolute, ahistorical concept expressed in Absolute Knowing. Although Habermas rightly acknowledges Hegel for his contributions to decentralizing the subject by introducing the means of language, labor, and mutual recognition as fundamental elements of his epistemology, he contends that Hegel's perspective still tends to preserve ahistorical mentalism (see Habermas 1999, 186).[34] According to Habermas, Hegel's ahistoric mentalism separates him from the later traditions of pragmatism, historicism, linguistic philosophy, and contextualism (see Habermas

---

33 Habermas 1999, 185.

34 By 'mentalism', Habermas means the paradigm of consciousness.

1999, 186–187). In particular, Habermas argues that Hegel denies the historical consequence of objective spirit by replacing it with a history of rationality that ends in Absolute Spirit (see Habermas 1999, 188). With the concept of Absolute Spirit, Hegel, according to Habermas, remains in the metaphysical structure of self-consciousness, and thus proposes an atemporal, ultimately mentalist framework.

Habermas objections against Brandom by virtue of his atemporal interpretation of Hegel mark the major turning point in the debate. It connects Brandom's approach directly to Hegel. However, Hegel's position is misconstrued. I will analyze Habermas' position on Hegel and demonstrate what I perceive to be the major mistake in his interpretation.

2.1 *Hegel's Denial of Mentalism*

For Habermas, Hegel develops a theory of practice that puts the family into the focus. By this, he reconstructs how human beings achieve common models of communication. He demonstrates how they produce a shared world. Unlike Brandom's formal semanticist approach, the reconstruction of the communicational model is based on a historical analysis. Habermas acknowledges Hegel's achievements in this regard, however, criticizes his ultimate result as ahistorical.

The main problem for the tradition of Hegel's time was the perceived difference between subjects and objects. This gap between subject and object was supposed to be bridged by different theories of self-consciousness (see Habermas 1999, 189). Posited beyond naturalist language and beyond presupposed principles of morality, the self-reflection of self-consciousness transformed the ancient and ethical imperative of "know thyself" into an epistemic task (see Habermas 1999, 190). The mental was consequently defined within these epistemic boundaries. Habermas identifies three problems that had to be addressed with regard to this problematic tradition: firstly, the myth of the given, secondly, the foundation of knowledge through the generation of knowledge in a subject, and thirdly, the equation of truth with certainty (see Habermas 1999, 190). In relation to these three problems, we encounter three dualisms: the dualism between I and not-I (something that is given to the subject as external), the dualism of inner and outer self (the private subject and its public generation of knowledge through language), and the dualism between immediate givens (immediate certainty) and mediated givens (see Habermas 1999, 190). In order to address all of these three problems and their resulting dualisms, Hegel, according to Habermas, had to detranscendentalize the subject.

Habermas first focuses on the Kantian origin of these problems: though Kant already developed the roots of a theory of self-consciousness, his approach, based ultimately on the transcendental unity of apperception, could not grasp the essence of subjectivity. Instead, it only objectified the subject. He identifies Kant's major flaw: the subject is affected by the empirical world, while, at the same time, Kant assumes a direct connection to mind-independent reality (see Habermas 1999, 191). This leads to a conflict between causal dependence and transcendental freedom (see Habermas 1999, 191). According to Habermas, it is the false mental paradigm that is the ground for Kant's problematic, dualistic answers (see Habermas 1999, 193). Ultimately, Kant cannot close the gap between experience and mind-independent reality.

For Habermas, Kant's project faces the problem that it cannot explain a mind-independent reality that affects us, but, at the same time, escapes our conceptual grasp. He believes that Hegel addresses this problem: "Hegel flatly denies that the knowing, speaking, or acting subject has to bridge an original gap between itself and the 'other'" (Habermas 2003, 182).[35] Hegel's solution is not to postulate a difference between the subject and its mind-independent reality. The solution is to investigate the relation of subjects to otherness. Since subjects are with 'others', the other can be grasped in terms of inter-subjective relations. This, according to Habermas, is the main reason why Hegel discusses the means [Medien] that make this sort of being-in-between possible (see Habermas 1999, 195). In the process of finding a solution for the problem of the ontological gap between mind-independent reality and the subject, the terms of subject and object will be replaced by the concept of objective spirit (see Habermas 1999, 196).

For Habermas, this fundamental change initiated in Hegel's theory is related to the traditional development of the social sciences [Geisteswissenschaften] and their historicist tendencies. Accordingly, philosophers were not only challenged to explain an eternal subject with regard to finite, natural objects but also with regard to their own historical formation in a possibly alienating society (see Habermas 1999, 196). Habermas argues further that for this reason, Hegel's description of human cultural development is shaped as a learning process, and not simply posited. In other words, all of our institutions are the historical results of social or natural problems. From here, particularly, labor [Arbeit] and language [Sprache] have determined the becoming of human society. Habermas writes with regard to these means [Medien]: "Language and labor provide media [means][36] in which the internal and external aspects,

35 Habermas 1999, 195.

36 I replaced the German word '*Medien*' with the term 'means.'

sundered by the mentalist approach, now merge" (Habermas 2003, 184–185).[37] This means that these forms of language and labor externalize the inner subjectivity of subjects, and their spirits persist while the subjects who originally generated them can vanish in the course of an emerging history. Humans, different from animals, do not only represent their species as individuals; rather, humans build identities and gain their personality within history (see Habermas 1999, 199).

The historical identities of individuals transcend the individual lifespan of only one individual. This has two consequences: firstly, we are not only observing humans as objects amongst other objects, but we need to acknowledge that each person has a history of growing into his personality (see Habermas 1999, 199). Secondly, this growth supersedes the individual person. Humans are more than merely learning individuals of their species. They are now part of a universal spirit of society. The question remains whether this universal spirit and the individual can be reconciled or whether both of them fall apart in society. As a consequence, Habermas focuses on the historical development of the community and its historical spirit.

As I noted above, for Hegel, the means [Medien] supposedly play a mediating role between the subjects and objects of a society. Hegel's solution to the problem of how the individual relates to the universal structure of spirit is therefore realized in means [Medien]. One of these means [Medien] is the family. Language, the theoretical side of our activities or the discourse of one single consciousness with nature, and labor [Arbeit] come together in a single society, while society's core is the family. As our lifebackground, these structures ultimately determine how we perceive our world (see Habermas 1999, 202). Through them, we no longer live as mere members of a species. Rather, we exist as members of a family in a single shared world. We have a particular relation to the world as members of this species. Habermas argues: "Because my linguistic knowledge organizes my actual perception, I cannot perceive anything without integrating it in a conceptual network" (Habermas 2003, 188).[38] This means that though 'reality' accompanies the practices of the participants, it is never experienced independently from the medium of communication. The concept 'reality' depends on practice, and more precisely on societal formations such as families.

As the last chapters of *Truth and Justification* express a shift in Habermas' theory towards relativism, Habermas finds a relativist criterion for objectivity

37 Habermas 1999, 198.

38 Habermas 1999, 202.

in the way we share spirit: "The collective spirit embodied in a community is as much 'objective' as it is intersubjectively shared by members who live from the same traditions and participate in the same practices" (Habermas 2003, 190).[39] In other words, objectivity is achieved through intersubjective practice. So, it is not only practical knowledge that produces institutions and tools. The standpoint of utility is superseded. In order to achieve objectivity, forms of mutual recognition like family structures have to come into play. Otherwise, Habermas believes, the overall structure of human interaction could not lift itself to the next level of cooperation.

How is the concept of family appropriately understood? For the most part, Hegel uses romantic love as the prototype of mutual recognition. In romantic love, it seems, the difference and unity of the subject and object can be achieved. Both lovers are different and one at the same time, leading to a relationship in which universality and individuality are reconciled in a unity. This union, however, lacks a stable particularity and can only be stabilized in a historical form of the family (see Habermas 1999, 205). Family is then an objective form of spirit that serves individuals to mediate their conflicts and to achieve cooperation. In love, the loving partners are worldless (see Habermas 1999, 209). They find an ideality that can never be achieved. In the family, however, the subjects are confronted with the question of an objective world because the relationship sustains even though partners will disagree. Their struggle for recognition is not completed by a single act of romantic affection, but sublated [aufgehoben] in the family.

In order to describe this sublated romantic relationship, Habermas refers to Hegel's theory of recognition and quotes: "Self-consciousness is in and for itself in so far as, and by virtue of the fact that it is for another which is in and for itself; that is, it exists only as something recognized" (Habermas 2003, 194).[40] So the different spirits that we will encounter must be results of the struggles for recognition, according through which each side fights to become an objective identity. With these struggles, we enter an epistemic quality of self-consciousnesses because it is relevant for self-consciousness to fight for a fundamental arrangement of how the world is. For Habermas, it is significant that the relationship of a family, for example, needs the presupposition [Unterstellung] of a shared, common world:

39 Habermas 1999, 204.

40 Habermas 1999, 209.

> Participants who find themselves related to one another in an intersubjectively shared lifeworld must at the same time presuppose [...] an independent world of objects [...]. A view cannot be shared if it is not a view of or about something obtaining in 'the' world, and a practice cannot be performed in common if it is not situated in what obtains in 'the' world, meaning that it is one and the same world for everybody.
>
> HABERMAS 2003, 193 [41]

It is this coordination of different perspectives which underlies true self-consciousness (see Habermas 1999, 211). The objective standard is thus neither mine nor yours (Habermas 1999, 212). Rather, it is a procedurally derived and shared view of how the world is.

In other words, Habermas moves away from his former conception of truth as justifiability and moves towards a conception of 'a world for consciousnesses' that is also for Hegel an important epistemological criterion. The true world is a world that we produce socially and historically. However, I will demonstrate that the Hegelian criterion is not inherently assumed as a prerequisite for effective communication. By reducing real human beings to mere subjects of communication, this criterion undermines their agency. While Habermas acknowledges the necessity of an objective world to mediate claims of knowledge, he views it as a product of practical discourse, in contrast to Brandom's perspective. According to Hegel, this criterion is not arbitrarily imposed but emerges as an inherent aspect of self-consciousness. It is an internal rather than external factor. In his rejection of the mentalist paradigm, Habermas solely relies on an empirical account of how the concept of 'world'

41 For Habermas, Pinkard captures this idea best: "The activities of making knowledge-claims is part of our overall practice of dealing with the world so as to satisfy the system of desires that make up our various projects; and we know things by integrating our conceptions of what counts as an authoritative reason for believing or acting into these overall projects and desires. Since two points of view can clash, there will be problems of conciliating one individual's claims with the conflicting claims of others. But a genuine conciliation could come about only if the parties could assume an objective, impersonal point of view [...] that is, only if they could judge their own claims not completely internally to their own point of view and experience but could judge them in terms of something that would transcend that subjective experience [...] Since the objective, impersonal point of view cannot be discovered [...] the agents themselves must construct a social point of view" ("Als Angehörige einer intersubjektiv geteilten Lebenswelt müssen sie unterstellen [...], daß es eine *identische* Welt von *unabhängig existierenden* Gegenständen gibt. [...] Eine Meinung könnte nicht mit dem normativen Anspruch verbunden werden, von allen »geteilt« zu werden, wenn sie nicht präsumtiv eine Ansicht von oder über etwas in der *objektiven* Welt ausdrücken würde") (Habermas 1999, 208; Pinkard 1994, 57).

is introduced in discourse. Consequently, he avoids engaging in an ontological exploration of the concept and treats it as an assumption within the discourse. It is at this point that Habermas' problematic interpretation and rejection of Hegel's concept of self-consciousness come into play.

2.2 *Habermas' Critique of Hegel's Concept of Self-Consciousness*

In order to further discuss Habermas' rejection of the 'ahistoric' conception of Absolute Spirit, we need to follow Habermas' view on Hegel's theory of the different stages of self-consciousness. Habermas begins his discussion with unhappy consciousness that, as an unstable form of self-consciousness, cannot find an ahistorical balance so that it exposes itself to doubt (see Habermas 1999, 212). His reconstruction of the following different stages of self-consciousness is brief. The doubt is replaced by a general form of reason that provides order for a better society. Enlightenment is supposed to recognize the order of nature and the order of society so that it can stabilize the unsettling nature of self-consciousnesses. On this aspect, Habermas writes: "the skepsis in skepticism finally gives way to the assumption of a common human reason that can justify, by its own devices, both an objectifying science of nature and an enlightened mode of ordering social life" (Habermas 2003, 196).[42] The problem for Habermas is that the historically generated human spirit could never fit into the picture of man that was designed during this project of enlightenment. Enlightenment just did not fully comprehend the lives of real human beings. Habermas remarks for that reason: "The essence of man escapes any 'scientific image of man'" (Habermas 2003, 196).[43] Because of these complications involved in theoretically capturing the true essence of human beings, the question remains about how to justify the structures of society. The gained theoretical position is not enough to support practically involved individuals.

At this point, Habermas believes that Hegel rejects the explanations of naturalism, historicism, and transcendental philosophy (see Habermas 1999, 214). Hegel, indeed, does not believe that intellectual culture can be explained naturalistically or by virtue of an entirely reflective transcendental project (see Habermas 1999, 213). Yet, instead of considering the option of Hegel's historicism, Habermas immediately argues that Hegel opts for an ethical self-clarification [ethische Selbstverständigung] (Habermas 1999, 213). So Hegel, according to Habermas, believes in a culture that assumes the possibility of self-grounding (see Habermas 1999, 213). The false and naïve belief

42 Habermas 1999, 213.

43 Habermas 1999, 213.

in self-grounding would therefore represent Hegel's ahistorical tendencies. Habermas criticizes that we cannot yet find universal forms that are capable of achieving this ahistorical self-clarification since they are still subordinated to the historical context. Thus, historicist contextualism would remain: "If they were, however, just part of a particular form of life, no genetic account of how people like us [...] could save our claims to validity from the suspicion of being as context-dependent as those superseded standards" (Habermas 2003, 199).[44] Habermas sees Hegel as an ahistoric foundationalist. The difficult point, however, is to determine the correct role that Absolute Spirit has in this process. Absolute Spirit, according to my historicist interpretation, is not the end of this process, but an end of epistemology. It is the entrance to categorical thinking that is timeless, yet historical. For all the following arguments, we have to keep in mind: *this timelessness does not mean that it is ahistorical.* Actually, it only means that *the representational understanding of time is insufficient to grasp the moments of spirit.* Hegel's project is grounded in the idea of eliminating all unnecessary presuppositions to clarify the preconditions of our thought. Time itself is an abstract thought that has to be sublated.[45]

Instead of considering Hegel's historicism in more detail, Habermas is suspicious of Hegel's Absolute Spirit: "whose is the 'spirit'" (Habermas 2003, 199)?[46] According to Habermas' suspicion, it is not the individual, but an anonymous

44 "Wenn sie aber intern mit dieser Lebensform verknüpft sind, ist es keineswegs ausgemacht, daß sie eine über diesen Kontext hinausreichende, universale Geltung beanspruchen dürfen" (Habermas 1999, 216).

45 The thought of 'sublation' even applies to Hegel's seemingly ahistorical logic. Indeed, the *Logic* will provide the oneness of the categories and also allow for a concept of externality. Historically, however, it is only the greatest compression of thoughts that we can achieve in our current historical moment. It is a diamond, compressed by the pressures and demands of our history. Thus, it is freed from the representational understanding of time, but it is not an ahistorical structure or framework. Given the instruments of cognition available to humans, it is our logic. In this sense, the problems we imposed on ourselves are freely chosen by the means of what we are as self-makers. Knowing can be acknowledged as absolute if we clarify all the factors of impact on the historically constituted subject. In order to give this historical interpretation of Hegel's *Logic* some plausibility, we only have to remind ourselves that Hegel claims in the preface of the *Logic* that he should have written the *Logic* 77 times, given that Plato revised *The Republic* seven times. Moreover, the introductions to the *Encyclopedia Logic* reveal that Hegel intended the *Logic* as a guide, yet not as a ready-made result. For Hegel, thinking trumps thought. With this in mind, it is true that we will be introduced to a sublation of time, but this does not mean that the following generated ideas are ahistorical. They have to be rethought in a constant process of historical thinking. I will clarify this in greater detail in my concrete discussion of Hegel's idea of Absolute Knowing.

46 Habermas 1999, 216.

form that takes over and governs our society. Habermas argues, however, that an ahistorical, all-encompassing spirit would not be *our* spirit (see Habermas 1999, 217). He argues against the ahistoricity of Absolute Spirit: "What is rationally acceptable by our lights is not necessarily what is objectively true. Even if it is guided by the idea of unconditional validity, finite spirit remains caught in its present and past and is thus ignorant of future and better knowledge" (Habermas 2003, 199).[47] In other words, Habermas believes that all our thinking that might achieve a higher level of rationality, can never justify itself as absolute, since it remains naïve with respect to the knowledge we will acquire in the future. Moreover, Habermas denies the possibility of a context of all contexts (see Habermas 1999, 219). He thus also rejects that we can achieve a final standpoint in the debate. History is open.

I do not object to Habermas' relativist views here. Nevertheless, Hegel's concept of Absolute Spirit is not related to a final standpoint. Obviously, Habermas takes Absolute Spirit as overcoming the differences between the objective and intersubjective realm so that Absolute Spirit can eventually guide the individual in order to get closer to reality. For Habermas, Hegel sees in the self-conceptualizing thought the Absolute Spirit that comes to itself as the substance of the subject. Habermas claims that Hegel does this by virtue of subjectivity, and thus he is not only an anti-historicist but also falls back into mentalism (see Habermas 1999, 221). It is no longer the mediation of language, labor, and interaction that guides subjects. Instead, the overall substance of everything that we do is the substance of this anonymous Absolute Spirit. So Habermas writes on Hegel's mental subject: "This subject is thought of as the One and All, as the totality that 'can have nothing outside itself'" (Habermas 2003, 204).[48] This totalized subject represents then a return to mentalism, which, for Habermas, constitutes the ultimately opaque precondition for Hegel's epistemology.

Though Habermas is right to critically examine the question of subjectivity, the most problematic part of Habermas' interpretation is that he, in my view, misunderstands Hegel's historicism. In the following, Habermas quotes a well-known, but often misinterpreted passage, in which Hegel argues that we have to eliminate our representational understanding of spirit's occurrence in time:

> Hegel identifies this 'other' with what happens in time, with the working out of the movement of the Concept, whereas the absolute self is

---

47 Habermas 1999, 219.

48 "Dieses Subjekt wird als das Ein und Alles, als Totalität gedacht, die »nichts außer sich haben kann«" (Habermas 1999, 222).

> understood as the Concept that engulfs time within itself, that consumes it, as it were. But this means that the historicity of reason ceases to pose a challenge. At the least, the challenge is blunted if logic once again wins out over history in the traditional way: 'Time is the concept which is there ...; this is why spirit necessarily appears in time, [but] it [only] appears in time for as long as it has not comprehended its pure concept, in other words not abolished time'.
>
> HABERMAS 2003, 204[49]

The problem is that Hegel has an idea of Absolute Spirit that must be thought of as beyond time, but this does not mean that it is outside of history. We thus need another conception of time and history that Habermas does not develop here or even consider. Habermas' problematic interpretation of Absolute Knowing as ahistorical instead of timeless guides his further evaluation:

> But the intersubjective structure of reciprocal recognition is no longer relevant for the mentalistic account of self and self-reflection, given that in the Logic Hegel unfolds the concept in accordance with the model of the 'ego', or of pure self-consciousness. Intersubjectivity is repressed from subjectivity, leaving no trace in the presentation of the absolute Idea.
>
> HABERMAS 2003, 205[50]

Because of his "Vereinigungsphilosophie" (a philosophy that unites), Hegel replaces, according to Habermas, objective spirit with a subjective idea of Absolute Spirit (Habermas 2003, 205). Thus, Hegel returns to the, for Habermas, forbidden mentalist paradigm that relies on ahistoricity. I hold that such interpretations inhibit a return to Hegel and have served Habermas to reject such

---

49 Hegel identifiziert dieses »Andere« mit dem Geschehen in der Zeit, worin sich die Bewegung des Begriffs vollzieht, während er das absolute Selbst als den Begriff begreift, der die Zeit in sich verschlingt und gewissermaßen verzehrt. Damit verliert die Historizität der Vernunft ihren herausforderenden Charakter. Sie wird jedenfalls entschärft, wenn die Logik über die Geschichte doch wieder auf die traditionelle Weise den Sieg davonträgt: „Die Zeit ist der Begriff, der da ist ...; deswegen erscheint der Geist notwendig in der Zeit, (aber) er erscheint (nur) solange in der Zeit, als er nicht seinen reinen Begriff erfaßt, d. h. nicht die Zeit *tilgt*" (Habermas 1999, 222). I will use Miller's translation of the *Phenomenology*: "Time is the Notion itself that *is there* and which presents itself to consciousness as empty intuition; for this reason, Spirit necessarily appears in Time, and it appears in Time just so long as it has not *grasped* its pure Notion, i.e. has not annulled Time" (Hegel PoS §801, 487).

50 Habermas 1999, 223.

a return. Though Habermas successfully refutes Brandom's model and criticizes his mind-external realism, he cannot successfully take up the threads of Hegel's historicism that ground our historical situation.

A reading of Hegel has to be established that grasps him as a historicist thinker who constructs an absolute that is open towards history. From here, the debate can be reinvigorated, instead of following further 20 years into understanding Hegel as a conceptual realist. In order to show that this is possible, I will reconstruct Hegel's epistemology as it is developed in the *Phenomenology* and interpret his passages on *Absolute Knowing.* Only then will we see why Hegel's and Brandom's philosophies are ultimately incompatible and why Hegel's philosophy is historically more viable in the long run.

CHAPTER 4

Hegel's Historical Relativism

In this chapter, I will argue that Hegel's concept of Absolute Knowing suggests that all cognitive claims are inherently tied to the historical moment. The notion of Absolute Knowing, as a cognitive claim about the identity of subjective knowing and objective truth, must be reconstructed as historical. For this reconstruction of Absolute Knowing as historical, I will focus on Hegel's distinction of time and history, introduced in the last chapter of the *Phenomenology of Spirit*. The *Phenomenology*'s historical dialectic that reaches its epistemological goal in the historicity of Absolute Knowing proves to be incompatible with Brandom's conceptual realism. Consequently, Rorty's claim that Brandom transitions analytic philosophy into its Hegelian stage becomes more challenging to defend. Based on my interpretation, both Hegel and Brandom offer competing models for constructing a viable epistemology, while I believe that Hegel is the model that we should adopt.

The previous discussion of Brandom's conceptual realism will structure my discussion of Hegel's historicism. There are three concepts that make Brandom's conceptual realism incompatible with Hegel's position:

1. Brandom's metaphysical realism as opposed to Hegel's relativist historicism,
2. Brandom's formal[1] game of giving and asking for reasons, and
3. Brandom's ahistorical semantics.

My discussion of these aspects will focus on *The Phenomenology of Spirit*. There are several reasons for this. First, Brandom relies almost solely on the *Phenomenology* as he takes it to be Hegel's central work and even goes so far as to reject other works of Hegel, such as the *Science of Logic*. Second, in the course of the surrounding discussion, Habermas has compared Brandom's approach mainly with Hegel's account from the *Phenomenology* and criticized both Brandom and Hegel for their ahistorical approach. Third, Rockmore, who I follow in interpreting Hegel as an empirical realist, sees the *Phenomenology* as introducing and presenting Hegel's central epistemological account, i.e., a historicist philosophy.

1 When I use the term 'formal', I am referring to the simplification of human language to a model of exchanging reasons through propositions contained in assertions, as discussed earlier. It is essential to note that Brandom's model also encompasses material inferences, which go beyond a purely formal logic.

 | DOI:10.1163/9789004714809_005

Before I reconstruct Hegel's historicism as I see it emerging in his *Phenomenology,* I need to address the writing of Hegel that, besides the *Science of Logic*, seemingly poses the strongest opposition to my historical and relativist interpretation. I will first review a discussion between McDowell, Sedgwick and Siep revolving around the question of how much Hegel was a historicist. Following Sedgwick and Siep, I hold the view that Hegel's *Science of Logic* should not be seen as the ahistoric center piece of Hegel's epistemology. In light of their discussion, I will then interpret Hegel's claim from the *Lectures on the History of Philosophy.* Hegel claims there that the historical development of philosophy mirrors the categories of an 'eternal' logic, which, however, contrasts with the method that he later applies in order to study history.

After these preparatory remarks, I will shift my attention to Hegel's *Phenomenology*, where I will illustrate that Hegel rejects the notion of content being separate from form. Consequently, defending the presupposition of progress toward a mind-independent reality becomes untenable, as the subjective form of cognition is inherently intertwined with its content. Therefore, recognizing a mind-independent reality equates to acknowledging a mind-dependent mind-independent reality. In other words, since all epistemology relies on a subject, the concept of pure content becomes problematic.

Secondly, I will demonstrate that Hegel's epistemology is based on a subject understood anthropologically and thus we cannot achieve a merely formal understanding of how we should act according to Brandom's conceptual realism and his normative pragmatics. Rather, our moral standards depend on the conditions of experiencing, social human beings in their historical context. These standards may change.

Thirdly, the reason for why Brandom believes he can relate to Hegelianism is based on a common conception that Hegel follows an ahistorical, social model of how we secure meaning by inferences. Habermas also perpetuates this, in my mind, problematic interpretation of Hegel as a social thinker, while ultimately rejecting Hegel's account as ahistoric. On the contrary, I intend to demonstrate that Hegel's account is intrinsically historicist and that accordingly all our cognitive claims are limited to the historical moment.

1 Logic and History

Houlgate finishes his review of Brandom's book with a negative note: "But, to my mind, it presents a reading of Hegel's *Phenomenology* in which, for the most part, one cannot place one's trust" (Houlgate 2020). One of his major criticisms is that Brandom "does not do justice to the project of the *Phenomenology* and its relation to speculative logic" (Houlgate 2020). He implies that the *Phenomenology* is merely Hegel's 'sceptical ladder' leading to the *Science of Logic* (see Houlgate 2020). Indeed, Brandom focuses on the empirical aspects of the *Phenomenology* and mainly develops an epistemology from it. He even admits that Hegel's *Logic* never made sense to him:

> I simply couldn't make anything of that work [Hegel's *Logic*]. In particular, I couldn't see how someone who understood everything I took Hegel to have understood in the *Phenomenology* about the nature of conceptual content and concept-use could be moved to go on to write that later work about whatever it was about.
>
> BRANDOM 2019, 764

In this regard, my reading of the *Phenomenology* coincides with Brandom's interpretation since I also see Hegel's major contribution to current epistemological problems in his phenomenological approach to experience. Regarding Hegel's *Logic,* Brandom therefore remarks that in dealing with "these exalted categorial and speculative-logical philosophical concepts" we treat "determinate concepts ... [as their] poor cousins" (Brandom 2019, 5). However, the "cash value" of the categories lies in what they "teach us about ordinary ground-level empirical and practical concepts" (Brandom 2019, 6). For the same reason, Brandom identifies Hegel's "master strategy" as "semantic descent", the question of how "metaconcepts" support our interaction with the empirical world in terms of "ground-level concepts" (Brandom 2019, 676). So instead of the *Phenomenology* leading to the *Logic* which is, indeed, a view that Hegel supports in the *Science of Logic*, there are reasons for why the *Logic* is tied back to the *Phenomenology* and only meaningful with regard to experience. As Jaeschke, for example, often points out, the *Logic* is the foundation, but only the foundation (see, for example, Jaeschke 2020, 7).[2]

Metaphysical concepts are not rejected in Hegel's systematic account of philosophy but must be understood from the standpoint of real experiencing human beings. Brandom argues that this view of Hegel's philosophy was forgotten since he had not left any first-rate students. As a consequence, Neo-Kantians interpreted Hegel's Geist "in neo-Cartesian terms, to yield a bizarre picture of a supersubject" (2019, 764, 765). As Jaeschke discovered and presented in his editorial remarks on Hegel's early religious writings, some of Hegel's texts have even been systematically destroyed to shift the perception of

2 Jaeschke, Walter. *Hegels Philosophie.* Hamburg: Meiner 2020.

Hegel as a 'pantheist' to a Christian theologian (see 2020, 31). Such misleading interpretations of Hegel may have, as argued before, set the path for Russell's explicitly anti-Hegelian philosophy.

Rejecting what Hegel argues in his debatable *Logic*, Brandom's central 'pragmatist' conviction is that there "can be no finally adequate set of determinate concepts" (Brandom 2019, 607). Rather, Brandom believes that the relativity of our cognition will be the ground for a healed post-postmodern society. He writes in his *Spirit of Trust* on the importance of relativity:

> that makes it possible to say [...] of the one who confesses that he did not get things right, that what things were for him is not what they are in themselves and [...] that there is no such thing as getting things (stably, permanently, finally) right.
>
> BRANDOM 2019, 607

In other words, Brandom's take on ground-level experiences strengthens a relativist interpretation of Hegel that I share. Contrary to Brandom's view of Hegel as a "semantic theorist" (Brandom 2019, 4), however, I argue that Hegel's work should be entirely thought of in terms of empirically realist relativism. Nevertheless, as genuine tragedy is not about the conflict between right and wrong, but concerns the conflict between two rights—an aphorism that is often, but possibly incorrectly, attributed to Hegel—I do not say that Brandom's approach is worthless, of course. As Hegel argues in his *Lectures on the History of Philosophy,* the "history of philosophy" does not display "the calm addition of new treasures to those already obtained" (my translation, Hegel 1986, 28).[3] So not doubting that Brandom contributes to the debate, I think the central idea of Hegel's philosophy is the opposite concept of history and not mind-external reality as the ground for our philosophical self-understanding. My thesis is that Brandom's idea of a mind-independent reality is the dialectical opposite of a skepticism that denies truth overall and should be sublated by a dialectical understanding of how our experience is transformed into something that can count as relative knowledge.

Depending on the traditional line of interpretation one follows, at least two of my claims can be seen as controversial. First, as already indicated before, I will, contrary to Brandom's conceptual realism, reconstruct Hegel as a historical relativist, which, in Hegel's case, is "tertiary" empirical realism (see

3 Hegel, G.W.F. "Vorlesungen über die Geschichte der Philosophie I." *Gesammelte Werke 18.* Edited by Eva Moldenhauer und Karl Markus Michel. Frankfurt: Suhrkamp 1986.

Rockmore 1997, 197) that, however, is not limited to experience. Second, I see Hegel not as working towards an unchangeable, eternal logic, an adamantine net, or a diamond of categories. In my view, Hegel brings us to a conception of reason whose center is our radical freedom of reflection within the boundaries of our historical becoming.

It is important to note that I am not alone in my interpretation. Rockmore, Siep, and Sedgwick push in the direction of Hegel as a historical relativist as well, even regarding the *Science of Logic*. As I have already argued in Chapter Two, Pinkard limits Hegel's approach as a mind-internal investigation in contrast to Brandom's mind-external realism.[4] Žižek regards Hegel as the philosopher who is most open towards history. And finally, it is Rorty who saw in Hegel's "temporalization of rationality" (Rorty 1982b, XLI) the important step away from metaphysics towards historicism (see Baumann 2019, 1).[5] Rorty noted that his historicism, however, instilled "fear and distrust" (Rorty 1982b, XLI) in the philosophical community. It may therefore be a relevant question of how much the "fear of relativism" (Rorty 1997b, 177) can allow for a discussion of Hegel's historicism. Beiser, for example, states that the "danger of historicism for the Enlightenment […] came from its implicit relativism" (Beiser 2005, 30).[6] According to Beiser, a 'dangerous' relativism claims that "all putative universal values are ultimately ethnocentric, invalid generalizations beyond our own specific time and place" (Beiser 2005, 30). He notes that without "laws […] valid for all times and places; the proper constitution for a people depends on its specific circumstances and history" (Beiser, 2005, 30). In light of these dangers, Beiser interprets Hegel's historicism as clearly 'refusing' "relativist consequences" (Beiser 2005, 30). In other words, Hegel, even though he is a historicist, must be an anti-relativist.

Since this view of Hegel as an anti-relativist is widespread, I will focus in the following part on Sedgwick and Siep who present a relativist interpretation of Hegel's supposedly most ahistorical work, the *Science of Logic*. Afterwards, I will discuss Hegel's concept of history as it occurs in the *Lectures on the History of Philosophy* and show how Hegel commits to relativism.

---

4 I do not include Pippin in this list. He distinguishes Hegel's pragmatic account explicitly from relativism and presents his *Logic* as metaphysical (as far as I understand his book *Hegel's Realm of Shadows*) in the sense of referring to an underlying essence of reality.

5 Baumann, Charlotte. "Hegel's realm of shadows: logic as metaphysics in the science of logic: by Robert Pippin." *British Journal for the History of Philosophy* 27 (6), 2019: 1256–1260. DOI: 10.1080/09608788.2019.1616157.

6 Beiser, Frederick. *Hegel.* New York, London: Routledge 2005.

1.1 *The Historicity of the Science of Logic*

In their contributions to Kreines' and Zuckert's book *Hegel on Philosophy of History* (2017), McDowell, Sedgwick and Siep discuss Hegel's position on historical relativity.[7] While McDowell, as another important representative of Pittsburgh's conceptual realism, argues for elements of Spirit being ahistorical and points particularly to Hegel's *Science of Logic,* Sedgwick sees even the *Logic* as embedded and depending on history. It is not an adamantine net, but changeable. Siep also argues against a 'viewpoint from nowhere' which would be necessary for timeless truths. However, he entertains the possibility of a 'reflecting reason' that allows for social criticism and goes prospectively beyond reason's historical limitations. In the following paragraphs, I discuss these three different interpretations regarding Hegel's relativist philosophy.

McDowell's opponent is Pippin and his idea of an independent "'self-making' of Geist" (2017, 19).[8] While Pippin claims that we become humans over "a historical time", McDowell holds that humans remain animals and are tied to their ahistorical nature (2017, 19). For example, although Spirit goes beyond its animal nature, the logical concept exists, as McDowell claims, "without ceasing to be the concept of an animal species" (2017, 18–19). In other words, Spirit is situated in an objective nature. In regard to Hegel's *Logic*, McDowell therefore also disagrees with Pippin's rather relativist position that all "procedures [...] are developmental [and] not deductive" (2017, 25).

From the standpoint of Pippin, Hegel's 'deductions' are "highly idealized, to the point of artificiality" or "historical reconstructions of actual developments" (Pippin 2008, 110).[9] Thereby, Hegel sacrifices the precision of former strictly, geometrical methods for the greater romantic ideal of aiming at a systematic whole. The *Logic*, which should have been rewritten 77 times, as Hegel describes it (Hegel 2010, 21), is not a completed and compressed diamond of unchangeable categories, but a historical improvement of former historical, logical considerations. McDowell objects to this interpretation of Pippin, arguing that although the internal unfolding of the logical concepts must be a more or less precise development, it is ultimately "not historical" but aims at the truth of ahistorical thinking (2017, 28). So, we improve the grasp of our thinking, as

7 Kreines, James and Rachel Zuckert. *Hegel on Philosophy in History.* Cambridge: Cambridge University Press. 2017.

8 McDowell, John. "Why does it matter to Hegel that Geist has a history?" *Hegel on Philosophy in History.* Edited by James Kreines and Rachel Zuckert. Cambridge: Cambridge University Press. 2017.

9 Pippin, Robert. *Hegel's Practical Philosophy: Rational Agency as Ethical Life.* Cambridge: University Press. 2008.

McDowell argues, but it is "not a series of steps from past understandings" to our current form of thinking (see 2017, 29). Rather, we historically discover the true essence of thinking. Hegel's final thought of the *Science of Logic* is then understood as the "self-externalization of the logical Idea" whereas thinking can adequately reflect on objective nature (2017, 29). McDowell thereby argues against Pippin that with the progression of the logical idea towards nature, Spirit is not "the Concept" giving merely "itself actuality" (2017, 29). Rather, Spirit is more and more accurately expressing an objective, as I understand McDowell, mind-external reality (2017, 29). Logic is thus understood as an objective essence of forms, whereas its goal is to make accurate and precise reconstructions of reality. As McDowell puts it, logic reveals the "forms that enable thought to pertain to objective reality" (2017, 29). He therefore objects to Pippin's historicist interpretation that "history serves for Hegel as the frame in which he offers legitimations for that supposed self-legislation" (McDowell 2017 25). Hegel's *Logic* is not a "theory of normativity" (McDowell 2017, 25). It is rather an objective fact that can describe reality, is discovered through history, and is not simply made within a historical context.

McDowell does not think that only the *Logic* is ahistorical, he also thinks that "the same is true for most of the *Phenomenology*" (McDowell 2017, 30). Except for the *Phenomenology's* chapter on Spirit, all sections before represent "progress" by revealing "how subjectivity relates to its objects" (McDowell 2017, 30). He thus argues in line with Brandom's view, who takes the *Phenomenology* as a 'story' about knowledge and its practicality in understanding a mind-external, ahistorical reality.

Contrary to McDowell, I read the *Phenomenology* as presenting a relativist theory of knowledge. Nevertheless, Hegel's concept of historicity hinges, indeed, on the question of how to interpret Hegel's *Logic,* which he himself considered the core of his philosophy. Since my goal in this book, however, is to point towards Hegel's historicity in the *Phenomenology*, I can include only some remarks on the *Logic.* Basically, I agree with Sedgwick[10] who does not interpret the categories of the *Logic* in terms of "permanent, unrevisable, timeless truths concerning the pure forms of thought, derived in a somehow presuppositionless manner" (Kreines 2017, 7). Sedgwick challenges the "standard picture" of Hegel's *Logic* (Sedgwick 2017, 34). According to her, it is false to reconstruct Hegel as a meditating Cartesian ego, uninterested in life, and "unaffected by contingency and change" (Sedgwick 2017, 35). Contrary to McDowell, Sedgwick

10 Sedgwick, Sally. "Remarks on History, Contingency, and Necessity in Hegel's Logic." *Hegel on Philosophy in History.* Edited by James Kreines and Rachel Zuckert. Cambridge: Cambridge University Press 2017.

therefore asks what it means that the *Logic* is part of the spiritual and not of the physical (see Sedgwick 2017, 35). It does not mean that we escape into an inner monologue that resembles a conversation of God with him or herself. Even though the *Logic* is an experiment "solely with [...] thoughts [...] in their complete abstraction" (Sedgwick 2017, 35)—rehearsing the premise of what it means to think purely—the logical forms are still intertwined with history. Sedgwick understands "Hegel's deduction" as interlinked with his rejection of reason as "the supreme self-legislator" (Pippin 1997, 162/168, see also Sedgwick 2017, 36).[11] She follows Pippin's position that the "Hegelian experiment" is very skeptical with regard to any last words (Pippin 1997, 163). The *Logic* is not the "bloodless dance of the categories" (848), or a "selfenclosed" (844) and "mere conceptual game" (Pippin 1990, 847).[12] Consequently, Sedgwick draws "no sharp line between the empirical and the a priori" in Hegel (Sedgwick 2017, 35). For example, in the 1812 Logic, Hegel refers to the *Phenomenology* as the deductive ground, whereas, according to Sedgwick's citation of Hegel, the Phenomenology "is nothing other than the deduction of the concept" (Sedgwick 2017, 41). For her, it is an indication that the *Logic* does not simply begin without presuppositions but is tied back to its emergence in experiencing human beings. Moreover, in all his published works, Hegel tells a "developmental story" that represents "a progression from less to supposedly more adequate systems—culminating, of course, in Hegel's own system" (Sedgwick 2017, 44). With regard to this progressive building of systems, the categories are idealized "products of Hegel's study of the history of ideas" (Sedgwick 2017, 28). As Sedgwick sees it with regard to Hegel's *Philosophy of History*, the logical forms are then the "abstract relatives" of the real developing, historical discourses (Sedgwick 2017, 28). So even though Hegel's philosophy idealizes certain forms, it is never an entire abstraction from the historical process. The historical process always takes place, even when a lonely philosopher meditates about logical forms (see Sedgwick 2017, 46). Despite the philosopher attempting to be unconcerned with contingencies, the philosopher can never fully deny the importance of a parallel history that accompanies his experiment with the ahistoricity of thoughts (see Sedgwick 201, 46).

Sedgwick acknowledges that our historical reflections are "engaged in a conceptually mediated or 'thoughtful' consideration of history" (Sedgwick 2017, 47). While history is still reconstructed as a necessary progression of reason, there is "no thinker [who] can stand outside her time" (Sedgwick 2017, 48). The

11 Pippin, Robert. *Idealism as Modernism.* Cambridge: Cambridge University Press. 1997.

12 Pippin, Robert. "Hegel and Category Theory." *The Review of Metaphysics.* 43 (4), 1990: 839–848. https://www.jstor.org/stable/20128948.

belief in an a priori development is then a "sheer fantasy" not acknowledging that hypotheses are fundamentally anchored in someone's cultural history and historical creativity (see Sedgwick 2017, 48).

Siep also raises doubts about the possibility of attaining a "view from nowhere" that is necessary for the existence of "timeless truths" within the framework of logical categories (Siep 2017, 50).[13] He understands Hegel's logical categories as developed a posteriori because pure logic would "contradict the [in the *Phenomenology* intended] overcoming of the form-content or concept-reality dualism" (Siep 2017, 59). Siep asks, however, how the philosopher could obtain a perspective on the future, even though he is subordinated to "the shapes of consciousness and its reversals during the history of experience" (2017, 51). He remarks that consciousness as "striving for knowledge about itself and the world" sets itself on a path of "permanent self-examination and self-correction" (Siep 2017, 53). During this challenge of self-exploration, there occur "periodic reversals of old, insufficient, and incoherent criteria and their replacement by new ones" (Siep 2017, 53). Philosophy as "the most fundamental reflection" can attempt to grasp the "epistemological and ontological implications" and "help with these revolutions" (Siep 2017, 53).

Siep indicates that the *Logic* plays a significant role for philosophy's 'assistance' in history. He remarks, however, that it is false to proclaim an "independent category theory" since the logical concepts are "abbreviations of the principal forms of self-conscious social activities" (Siep 2017 58–59). However, the annulment of time in the *Phenomenology* as a partial annulment of history possibly allows for an internal and higher standpoint. He thereby softens Hegel's relativist historicism: "This certainly does not mean that history as a temporal process comes to an end, but that for the comprehension of the concept [...] the historical process loses its significance" (Siep 2017, 61). Like Habermas, Siep overlooks here that Hegel speaks about time as a concept of intuition which does not include the concept of history. Hegel's concept of history, as I will show in the following parts, requires the 'annulment' of time in order to be appropriately grasped as only conceptual. The progression of time, since it is not an absolute that is or can be immediately perceived, can only be properly understood if we grasp it through the historical constitution of ourselves. Humans' experiential involvement in history, to put it quite Heideggerian, precedes the abstract concept of time. Siep, on the contrary, takes himself to be following Hegel when he argues that the "essential structure

13 Siep, Ludwig. "Philosophy and the 'Stream' of Cultural History." *Hegel on Philosophy in History.* Edited by James Kreines and Rachel Zuckert. Cambridge: Cambridge University Press. 2017.

("pure concept") of spirit is disclosed in speculative philosophy" including "a systematic sequence of determinate concepts" (Siep 2017, 61). Developing this thought, he then moves in the direction that the 'elaboration' of the categories can, for example, reveal the "basic institutions of a rational state" (Siep 2017, 61).

Despite these remarks on the predictive power of the *Logic,* Siep is careful not to draw far-reaching conclusions and reminds us that Hegel does not drift into a simple formalism of logical categories. Hegel's claim of the parallelism of the logical categories and the historical shapes remains, indeed, controversial, as I will also explain in the next point. Yet, Siep, despite his careful interpretation, believes that the *Logic* delivers an interpretative plateau from which we can evaluate our history. With the *Logic* "a level of conceptual norms is reached that allows judging the progressive direction of the past and the possible setback of future developments" (Siep 2017, 62). According to him, absolute knowledge can then supersede "the logic of (past) experience" (2017, 62). Hegel's philosophy is consequently not only retrospective, but delivers criteria for the evaluation of future events (see Siep 2017, 62).

I am not entirely opposed to Siep's proposals for a progressive evaluation of history as long as it is within the boundaries of historical relativism. Siep's search for a Hegelian safe-spot, however, seems to be grounded in the worries he expresses with regard to the diverse dangers of relativism. Claims for radical change, for example, as long as they are "supported by strong majorities" sacrifice philosophy's role of "social criticism" (Siep 2017, 64). Siep warns that "renouncing a priori principles" has "high costs for the philosopher" (see Siep 2017, 66). So, despite admitting that the philosopher kings are dethroned since there is no strong external viewpoint anymore, Siep urges us to preserve philosophy's strength of social criticism.

Nonetheless, because relativism implies certain inconveniences for philosopher kings and social critics, it does not constitute an argument for why we should return to ungrounded principles of the modernity project. As I perceive it, moreover, Hegel's philosophy can criticize and evaluate societies within a relativist historicism. I agree with Siep when he ultimately affirms a more historical position: "Without firm river banks, milestones, and compasses, this might mean using only slow-moving objects for orientation" (2017, 67). It is not about the universal, immovable standards that stand out from history. Rather, it is about standards that pass only the tests of our historical moment.

Regarding the current discussion, it becomes clear that I disagree with Brandom's overall edifying and ahistorical interpretation of Hegel as a prophet. I side, however, with Brandom's relativist position regarding the logical categories, a position that is also held by Hegel scholars such as Siep and Sedgwick.

The logical categories serve as a historical, preliminary ground to interpret ground-level phenomena. They are a historically changeable, practical device for understanding empirical phenomena in a historically grown society. It is necessary for humans to develop such 'slow moving' (Siep) objects, but it is not necessary that they are unchangeable. The question remains: what is history according to which these objects occur and in specific how does the *Phenomenology* prepare Hegel's concept of historicity?

1.2 *Hegel and the Concept of History*

After having clarified that even the *Logic* can be reconstructed within a historical relativism, I would like to discuss Hegel's direct thoughts on history with regard to the *Lectures on the History of Philosophy*. In that regard, it is important to investigate the controversial parallelism between logic and 'experienced reality' [*Wirklichkeit*]. In his *Lectures,* Hegel claims that history, at its core, develops in the same way as the logical categories are related to each other and unfold. Since this parallelism thesis can be read as one of the strongest objections against my relativist interpretation of Hegel, I will explain how Hegel came to this parallelism thesis and contextualize it.

Hegel held his *Lectures on the History of Philosophy* already 1805/06 in Jena and regularly taught this subject until his death in 1831 (see Jaeschke 1993, VIII).[14] In his preface to the lectures, Jaeschke, as the last editor of the collected works of Hegel, emphasizes that these lectures have a special status in the corpus of Hegel's works. No other subject has been taught as frequently and as extensively by him (see 1993, VIII). However, in that regard Jaeschke also addresses the challenge of composing a unified text from different documents (Jaeschke, 1993, XI–XIII). Given the coexistence of the parallelism thesis and other historical relativist theses, it is advisable to begin by examining Jaeschke's editorial remarks before delving into the text itself.

According to Jaeschke, Hegel pioneered a new interpretation of history that did not present history as only an aggregate of facts. He explored whether philosophy could provide a rational reconstruction of history centered around the evolving and unfolding of reason (Jaeschke, 1993, VIII). Regarding this role of reason, the *Lectures* mark a turning point in Hegel's view of history. Before 1805, Hegel believed that each genuine philosophical system would eventually attain absolute truth. He argued that "each [form of] reason [Vernunft], which directs itself to itself, produces a true philosophy" (Jaeschke 1993, X,

14 Jaeschke, Walter. "Einleitung." *Vorlesungen über die Geschichte der Philosophie—Einleitung, Orientalische Philosophie by G.W.F. Hegel.* Edited by Walter Jaeschke. Hamburg: Meiner. 1993.

my translation). Through the process of reflection, each philosophy holds the inherent capacity to surpass the limitations imposed by its historical context. It can reveal truth. However, Hegel underwent a revision of his perspective in his *Lectures*. He now believes that philosophical systems evolve, surpassing previous frameworks and progressing towards a greater whole (Jaeschke 1993, x).

In that regard, interpreting the claim of parallelism between historical and logical development in Hegel's lectures poses a significant challenge. Jaeschke questions whether history unfolds according to the same plan as logical categories (Jaeschke 1993, xviii). For example, finding accurate and appropriate logical concepts that align with the historical development of philosophies proves challenging, such as reconstructing Heraclitus as an Eleatic follower (Jaeschke 1993, xviii). Additionally, Hegel struggles to identify a historical position that adequately expresses the logical category of nothingness (Jaeschke 1993, xviii). Jaeschke concludes that a significant reduction in accuracy is required to interpret history based on the proposed parallelism (Jaeschke 1993, xviii). As an alternative, Jaeschke therefore explores the idea that the "real principle of order in Hegel's history of philosophy [...] is nothing but history itself" (1993, xxiii). This suggests that history, although logical, cannot be abstracted like logical categories but must be understood reflectively.

Regarding Hegel's applied method, there is no doubt that Hegel did not follow his programmatic parallelism thesis (see Jaeschke 1993, xxiv). Other attempts at reconstructing history according to the program of the introduction are therefore, according to Jaeschke, dogmatic and give the "impression of sterile Hegelese" [*Eindruck steriler Hegelei*] (1993, xxiv). They misunderstand Hegel's "real procedure" [*wirkliches Verfahren*] (Jaeschke 1993, xxiv). Yet, Jaeschke also remarks that this does not mean that there is no philosophy in history (1993, xxv). He notes: "Who refuses the logic-historical identity, does not need to feel obliged to claim the opposite extreme" (1993, xxvi).[15] Hegel is still interested in understanding history in terms of necessity (see Jaeschke 1993, xxvi). However, the process of history follows its own integrity that in its complexity cannot be simply abstracted and reduced to logical categories.

However, Jaeschke also wonders about Hegel's motivation to claim without proof that history and logic develop according to the same plan, especially since Hegel eventually follows another method of studying history (see 1993, xxvii). Jaeschke explains that Hegel wrote the lectures before the *Phenomenology* in

15 „Wer die logisch-historische Identität verwirft, braucht sich nicht zur Behauptung dieses entgegengesetzten Extrems nötigen lassen" (Jaeschke 1993, xxvi).

1805/06 and takes from there, as he can only assume, the parallelism thesis. It is the much later Hegel of Berlin who eventually arrives at a conception of history as independent of logical concepts (1993, XXVIII). From then on, it is about the "moments of mediation" [*Vermittlungsmomente*] between history and logic. As Jaeschke argues, Hegel does not replace the parallelism thesis as part of the programmatic introduction anymore. It may be for the sake of simplification given that it is a general introduction that, moreover, was often complemented by oral remarks (Jaeschke 1993, XXIX).

The parallelism thesis poses a significant challenge to a relativist interpretation of Hegel, as it implies a fixed and unalterable essence within logic. However, as Jaeschke argues, this thesis does not align convincingly with Hegel's notion of history. To support my historical, relativist interpretation of the *Phenomenology*, I will examine Hegel's introduction to the *Lectures on the History of Philosophy*. In the subsequent paragraphs, I aim to explore the role of the parallelism thesis within the context of my proposed relativist interpretation.

In his *Lectures on the History of Philosophy*, Hegel discusses the "inner contradiction" [innerer Widerstreit] of the history of philosophy (Hegel 1986, 24).[16] The contradiction concerns the idea that truth as philosophy's "aim" is "unchangeable" and therefore "has no history", while history seems to be about "temporary" objects emerging and disappearing within the contingent horizons of time (see Hegel 1986, 24). In order to resolve this contradiction, Hegel focuses on the *aim* of eternal truth that emerges within the historical context of human beings.

It is important to emphasize here that Hegel does not claim that we hold an eternal truth directly since it would remain an abstract thought considering the magnitude of the whole and the finiteness of human beings. Rather, he addresses in the *Lectures* how humans *aim* at such an eternal truth by reflecting on themselves within a historical development. The aim of that movement is Spirit's identity with itself in a specific historical moment. This identity with itself is embedded in history, achieved through history and represents something that can be called a procedural, eternal truth. In the following paragraphs, I will reconstruct how Hegel developed this thought of a procedural, eternal truth.

The first important opponent of Hegel's approach is a skepticism that subordinates philosophy to the contingencies of history and thus rejects a trajectory

16 Hegel, G.W.F. *Vorlesungen über die Geschichte der Philosophie—Einleitung, Orientalische Philosophie*. Edited by Walter Jaeschke. Hamburg: Meiner, 1993.

of eternal truth. Emblematic for this position is, as also referenced in the preface of the second part of the *Science of Logic,* Pilate asking Christ: "What is truth?" (see Hegel 1986, 32, see Hegel 2010, 507). According to Pilate's cynical skepticism, 'truth' is disregarded as an already overcome, obsolete position that is not worth consideration. Truth cannot be found as it is an empirical fact that great minds have erred and too many philosophies have contradicted each other. Hegel's argument against this skeptical position is simple. He points to the fact that all different, historically known philosophies have at least held their own position to be true (1986, 34). The argument appeals to what is known as the performative contradiction today. It means that even the skeptical philosophy resulting from the ongoing 'battle of metaphysics' must admit that it takes some kind of truth for its own position for granted. It means that the skeptic believes it is true that truth is obsolete or that nothing can be known.

Hegel does not fully develop these arguments, but the brevity of them indicates that he does not see such kind of cynical skepticism as a philosophically strong position. In the *Phenomenology of Spirit,* for example, he compares it with "the squabbling of self-willed children" who oppose everything the other party claims (PoS §205, 126). It is therefore important to remind ourselves that the presented skepticism is not the same as what I mean by relativism, which in its particular terminology was an unknown position at Hegel's historical moment. As I have distinguished in the introduction of this book, relativism allows for the aim of an eternal truth, while crude skepticism denies that truth is possible overall.

Although Hegel clearly positions himself against skepticism, he does not discuss it in further depth. He is rather concerned with a second opponent: philosophies that posit truth as an "immediate perception [or] intuition" (1986, 34). Such positions that constitute only "contingent thoughts" must be regarded as promoting merely opinions (Hegel 1986, 29), while philosophy seeks a "general thought for itself" (1986, 31). With regard to this aim, philosophy "is the objective science of truth, of necessity and concrete recognition" (1986, 31). Here, Hegel qualifies truth as a question of recognition that is not simply intuited or perceived but requires the "labor [Mühe] of thought" (1986, 34). As I interpret it, Hegel faces a significant undertaking that necessitates navigating a course that steers clear of both an all-encompassing skepticism that negates the very notion of truth and an intuitive philosophy that purports to possess truth in an enigmatic manner, detached from the arduous path of rigorous thinking. It means that even though philosophy is the "objective science of truth", Hegel's vision of philosophy remains within the boundaries of what I understand as relativism. We have to qualify what this relativism means with regard to the *Lectures* in the next paragraphs.

First of all, we will find several passages, in which Hegel seemingly speaks out against relativism, particularly, when he discusses the eternal aim of philosophy. One of the critical passages in which Hegel speaks seemingly against relativism is the following (since the standard translations tend to, in my mind, misinterpret its meaning, I will cite the following passage in German and translate it thereafter):

> Bei dieser Betrachtung gehen wir freilich davon aus, daß die Philosophie das Ziel habe, die Wahrheit denkend, begreifend zu erfassen, nicht dies zu erkennen, daß nichts zu erkennen sei, wenigstens daß die wahre Wahrheit nicht zu erkennen sei, sondern nur zeitliche, endliche Wahrheit (d.h. eine Wahrheit, die zugleich auch ein Nichtwahres ist); ferner, daß wir es in der Geschichte der Philosophie mit der Philosophie selbst zu tun haben.
>
> HEGEL 1986, 38

The passage focuses on our natural assumption that philosophy has an aim and rejects skepticism, i.e., the claim that we cannot know. The conventional translation omits certain nuances present in the original German text, thereby reducing ambiguities:

> This makes it easy to us to comprehend the aim of Philosophy, which is in thought and in conception to grasp the Truth, and not merely to discover that nothing can be known, or that at least temporal, finite truth, which also is an untruth, can alone be known and not the Truth indeed.
>
> HEGEL 1995, 19

I would like to suggest a translation that I perceive to be more accurate with regard to some ambiguities. The word 'freilich' is rather a modal particle and has therefore more of a pragmatic meaning. The best translation is not 'easily' but 'of course'. Moreover, the construction 'von etwas ausgehen' means 'to assume', and 'Betrachtung' can be translated with 'investigation'. Moreover, I substituted 'begreifend zu erfassen' with a conventional Hegelian term, namely, 'conceptual thinking'. The following translation may therefore be more concise:

> With regard to this investigation, we assume, of course, that philosophy has the aim to grasp truth through conceptual thinking.
>
> HEGEL 1986, 38

My translation emphasizes that we *assume* that philosophy aims at truth. It is important to note that this does not definitively establish whether Hegel believed that we will ultimately attain Truth or not.

The second part of the sentence presents difficulties, as the grammatical structure is uncommon even for native German speakers. Here is my suggested translation for the second part of the sentence after [it is the naturally assumed aim of philosophy]:

> to recognize not that nothing can be known, *at least* that the true truth cannot be known, but only temporal, finite truth (that means a truth, which, at the same time, is also a non-truth); further, that we recognize that, with regard to the history of philosophy, we deal with philosophy itself.
>
> HEGEL 1986, 38

Hegel's uncommon use of the word 'wenigstens' poses a major problem for an accurate translation. First, Hegel emphasizes that it is not the aim to recognize that nothing can be known. This part of the sentence is clear. Then, however, he uses the word 'wenigstens' which I translated with 'at least'. The positioning and ambiguity of this word hold significance, as they give rise to different interpretations.

A) The first option is to interpret this sentence introduced by 'wenigstens' as a complete denial of relativist truths. Then, it sounds, however, as if relative truths constitute the worse belief when compared with the belief that nothing can be known. This is already a bit absurd. Moreover, it is unclear what Hegel then means by the following 'further' [ferner]. The sentence could mean: we should 'to the least' assume that philosophy deals with contextual truths and 'further' [ferner] we should not assume that the history of philosophy is philosophy. Hegel, however, clearly holds the view that the "history of philosophy deals with philosophy itself" (1986, 38). Given these complications, we need to consider another option.

B) the second option entails interpreting Hegel's aim as the resolution of the conflict between contextual and eternal truths. This interpretation highlights the significance of the semantic pairing of 'at least' and 'further', emphasizing the need to acknowledge the coexistence of historical truths alongside the potentiality of eternal truths. By adopting this interpretation, we address the contradiction initially presented by Hegel, wherein he explores the notion of an 'eternal' philosophy having a historical dimension.

As already indicated before, a more genuine relativism allows for the possibility of an eternal truth, while we must walk the path of contextual truths, as long as we have not yet convincingly demonstrated such a truth. Moreover, we cannot simply intuit the eternal truth of philosophy, since it would remain abstract. So, we need to deal with the specific, historical conditions that precede our historical moment. In this sense, the sentence above supports rather a historical framework than a framework of an ahistorical truth.

In my interpretation, Hegel, as a philosopher striving for eternal truth, engages in the discussion of historical thinking and goes so far as to designate it as the essence of contemporary philosophy. This assertion is supported by Hegel's later statements, where he expresses:

> Each philosophy is philosophy of its time. It is a part in the complete [ganzen] chain of spiritual [geistigen] development. It can thus only satisfy the interests that are adequate to its time.
>
> HEGEL 1986, 65[17]

This passage, along with others, provides support for a relativist interpretation of Hegel's philosophy, despite the fact that each philosophy aims to attain eternal truth. Specifically, Hegel's aim is to bridge the pursuit of eternal truth in philosophy with the inherent limitations of human understanding within a historical context. This viewpoint is further reinforced by his earlier ideas. At the core of Hegel's philosophy is the reconciliation of the "opposition between Truth and Error" (1986, 37). Hegel presents the history of philosophy as a diverse tradition marked by conflicting positions, all driven by the shared objective of an abstract, eternal truth. To comprehend the relationship between historically bound human beings and eternal truths, I will argue in subsequent parts that Hegel transforms the traditional notion of truth. Traditionally, 'truth' is understood as the correspondence between subjective thought and an external reality. However, a problem arises: an "ontological hiatus" exists between the subject and the external reality, as also described by Brandom. Hegel seeks to resolve this problem through the concept of Spirit, which encompasses both subject and substance. Through a comprehensive historical process of reflection, individuals as subjects can gain awareness of their substance, attain the identity of Spirit, and establish their grounding within it. This implies that

17 "Jede Philosophie ist Philosophie ihrer Zeit, sie ist Glied in der ganzen Kette der geistigen Entwicklung; sie kann also nur Befriedigung für die Interessen gewähren, die ihrer Zeit angemessen sind" (Hegel 1986, 65).

humans, employing historical reasoning, can recognize and actualize their own truth, shaped by their historical development.

The project of humans' historical self-reflection—in order to achieve the identity of Spirit—constitutes a bidirectionality of philosophy. First, the "aim of philosophy" is thereby to "recognize truth" by deriving it from the historically occurring phenomena (1986, 39). Second, it is to understand all phenomena from the base of philosophy, in this case from self-reflection (see 1986, 39). The connection of both directions is the historically developed and in its history self-grounding reason of human beings. This envisioned self-grounding circularity of reason is achieved when Spirit as the objective history of thoughts and the Spirit as the subjectively actualized potential of human reason become the same. Hegel describes this as Spirit's self-identity and explicitly calls it the "absolute aim" (1986, 42). More specifically, he writes: "Spirit's being-with-itself, this coming-to-itself can be called its highest and absolute aim" (Hegel 1986, 41).[18] This absolute aim, as Hegel develops this notion further, is freedom (see 1986, 42). It is freedom because humans make themselves transparent to themselves by means of what they have historically become so that by virtue of their transparency they can overcome their dependence on anything outside of themselves. They understand why they are at their particular moment in history and why they historically act the way they act. Hegel also calls this achieved self-transparency 'authentic ownership' [wahrhaftes Eigentum] (1986, 42).

Now, the question remains as to how this identity, arising from self-reflection, can genuinely manifest for human beings within their historical context. In Hegel's terms, spirit as substance must achieve itself in subjective thought. The subjective thought through which Spirit attains freedom from all constraints is an identity that Hegel now refers to as the concrete. Reflecting on this concrete, however, we will see that it is composed of further two contradictory principles.

These two contradictory principles of the concrete are, first, development, which stands for the historical process towards the identity of spirit, and, second, rest, which stands for eternal truth (see 1983, 43). Ideally, the concrete as a totality of human thoughts embraces everything: everything that has been, is, and will be. In a sense, it eliminates the relevance of time and thus constitutes its eternal truth as timeless. However, because of their phenomenological limitations, finite human beings are not equipped to think everything at once. The

18 "Dies Beisichsein des Geistes, dies Zusichselbstkommen desselben kann als sein höchstes, absolutes Ziel ausgesprochen werden" (Hegel 1986, 42).

key to this contradiction for Hegel is therefore that the finite mind develops through thinking and keeps developing with regard to its eternal aim. The eternal aim of this developing mind is the concrete as an idealized *timeless* totality of thoughts. The aim of *timelessness*, however, remains abstract and cannot be reached by finite minds. Rather, the truth of the totality of thoughts has to be tested in the experienced history of human beings again and again. Humans can understand how the aim of the totality then constitutes their historical essence. As a self-produced aim, it directs their thoughts and includes the intent of eliminating the contingent aspect of time. Through this elimination of time, they identify with this aim of eternal truth. But this identity between the subject and its historical essence of striving for truth is not ahistorical, as, for example, argued by Habermas and others. In its elimination of time, it only strives to eliminate contingency at a certain historical moment. The basis for my interpretation of the *Phenomenology* and this passage is that timelessness and ahistoricitiy are different concepts.

We have two perspectives regarding the concrete: with regard to the history of philosophy, 'concrete' means that it develops as thinking (see Hegel 1986, 39). With regard to the question of the totality of thoughts, the 'concrete' rests as the eternal aim of the same thinking (see Hegel 1986, 39). As Hegel translates development and rest into Aristotelian terms, thinking humans actualize their potential. Development then means that something that is in-itself actualizes its potential and thus becomes for-itself (see Hegel 1986, 39). Later, Hegel declares that the movement of actualizing its potential has to be the central focus of philosophy: "The more this development is achieved, the more perfected is the philosophy" (1986, 47).[19] Philosophy is thus the question of how to perfect the development with regard to its internal aim, which is rest. Since rest is abstract, however, it can, so far as we know, only be realized through a constant movement in our current history (see Hegel 1986, 47). In this model, history never pauses, even though we strive to eliminate the attached contingencies of time.

Now, Hegel does not expound on the complicated structure of this paradoxical unity of rest in development, but he compares it to the unity of an organism (see 1986, 47). As I understand it in terms of the later Anthropological Philosophy without presupposing a telos: a living organism is a composition of constant, internal movements. Yet, even though there is no stable internal entity—because everything in it is movement—it does not fall apart in its

19 "Die Philosophie ist nun für sich das Erkennen dieser Entwicklung und ist als begreifendes Denken selbst diese denkende Entwicklung. Je weiter diese Entwicklung gediehen, desto vollkommener ist die Philosophie" (Hegel 1986, 47).

differences. The internal movements constitute a unity that appears as a body. Without its organized internal movement, the organism would be dead and decompose.

Hegel's approach to reconciling the contradiction between history and truth closely aligns with the concept of an organic body. The aim of historical, finite human beings who seek truth does not lie in progressing towards a static, external reality. Similar to an organism, their unifying aim is internally constituted, shaped by their historical aspirations as humans. Their aim is not teleologically outside of them; rather, they seek the truth of their own existence in societal and historical contexts. Unlike mere organisms, however, they exercise their unique capacity for self-reflection, which Hegel identifies as an expression of freedom.

Beiser holds a contrasting view regarding Hegel's concept of Spirit and its internal aim of freedom. When reflecting on the concept of the spirit of the people (Volksgeist), Beiser argues that Hegel took this central concept of the historicists and framed it in Aristotelian terms, thereby transforming it into the underlying "formal-final cause" of freedom which is shared by "all nations" (Beiser 2005, 212).[20] Beiser believes that this teleological composition of Spirit saves Hegel's historicism from the problem of relativism (see Beiser 2005, 212). Specifically, it addresses the issue of lacking a standpoint of orientation, which Beiser sees as a problematic skeptical position that rejects the pursuit of eternal truth. Beiser points out that the "self-awareness of freedom" as "the goal of world history [...] provides a single [...] criterion of value" (Beiser 2005, 212). Consequently, cultures can be evaluated based on their degree of realized freedom, which, according to Beiser, is their ultimate, mind-external goal (see Beiser 2005, 212). Freedom, however, transcends mere realization within societal bounds; it can be perceived even more radically as the recognition that humans resist reduction to their essence. Instead, they possess the potential to evolve into something that surpasses any previous limitations. Thus, freedom emerges as a concept that transcends its own boundaries, expressing the infinite capacity of human potential. In response to Beiser, moreover, it is crucial to acknowledge that Hegel can develop a relativist, historical criterion based on the metaphorical 'organic' movement he describes. Thus, the presence of a plausible criterion such as freedom for evaluating the development of societies does not negate its relativist nature. It is indeed true that Hegel presents a criterion for assessing cultures through a historical analysis of what has guided their progression. He arrives at a concept of freedom that

20 Beiser, Frederick 2005. *Hegel.* New York, London: Routledge.

can serve as a rational criterion. However, this does not automatically dismiss relativism. In fact, I argue that genuine relativism acknowledges the possibility of warranted criteria while recognizing that these criteria are contextual and not necessarily enduring truths. They represent truths of their specific historical moment. The criticism of relativism often stems from the misconception that relativists claim 'all moral systems are equal'. Nonetheless, certain forms of relativism can offer criteria for distinguishing moral systems, albeit without asserting the independence of these criteria from their historical context.

As Beiser observes, Hegel as a historicist is, indeed, also a critic of certain types of historicism, but, as I see it, not because he rejects relativism. He rather rejects a historicism that does not deliver any criteria, which includes historians that *aimlessly* record historical events. Criteria are necessary for orienting ourselves at our historical moment. At this point of the argument, Hegel therefore introduces his often discussed parallelism between history and logical concepts as a criterion for evaluating the progress of philosophy towards its aim. According to this parallelism, the history of philosophy is a mirror image of the system of logical concepts (see Hegel 1986, 48, or Hegel 1986, 61). In other words, all sequences of our historical development can be reconstructed according to the system of logical concepts:

> According to this idea, I now claim that the sequence of the philosophical systems in history is the same as the sequence as logical derivative of the conceptual determination of the idea.
>
> HEGEL 1986, 49[21]

If we did not seek these logical concepts in history, we would, as Hegel adds further below, only deal with "an unruly heap of opinions" [ein unordentlicher Haufen von Meinungen] (1986, 50). So, parallelism is a criterion for the sufficiency of a history of philosophy. However, as adamant as the parallelism claim sounds, Hegel immediately softens it as an ultimate criterion and admits that we do not see the logical laws directly in history. He then remarks that a deeper investigation of this parallelism would lead us too far away from

21 "Nach dieser Idee behaupte ich nun, daß die Geschichte der Philosophie | Aufeinanderfolge der Systeme der Philosophie in der Geschichte dieselbe ist als die Aufeinanderfolge in der logischen Ableitung der Begriffsbestimmungen der Idee. Ich behaupte, daß, wenn man die Grundbegriffe der in der Geschichte der Philosophie erschienenen Systeme rein dessen entkleidet, was ihre äußerliche Gestaltung, ihre Anwendung auf das Besondere und dergleichen betrifft, so erhält man die verschiedenen Stufen der Bestimmung der Idee selbst in ihrem logischen Begriffe" (Hegel 1986, 49).

our current discussion (see 1986, 49). Rorty would say that in this instance, Hegel—despite his project of the "temporalization of rationality"—had one of his "Philosophical moments", thereby criticizing Hegel for holding on to a dogmatism of philosophically secured "Truth" (Rorty 1982b, XLI). Yet, contrary to Rorty's view, I argue that Hegel holds a view of the logical categories as flexible and even historical. The parallelism can therefore be seen as another reflection on historical truths. If we do not take the *Logic* as an adamantine net, the parallelism thesis is way less controversial.

I believe that with the parallelism thesis, Hegel considers the need for relativist theories to produce criteria with an aim of truth, which in this case are the very useful, logical categories. Without specifying the parallelism, however, Hegel gave us the riddle of how to interpret the open question of the relationship between logic and our historical reality. In line with Jaeschke, I argue that Hegel's project of the ensuing necessity must be rather understood with regard to the real procedure [*wirkliches Verfahren*] that he uses in the *Lectures*. This means that Hegel thinks through the concrete history of philosophy and finds 'moments of mediation' between *Logic* and history (as Jaeschke calls it 'Vermittlungsmomente' [Jaeschke 1993, XXIX]).

The question remains how Hegel's 'real' method can be described. As he already prepared this method in his *Phenomenology of Spirit,* I claim that it is phenomenological. Despite its title, surprisingly few interpreters have asked whether Hegel's *Phenomenology* could actually be a phenomenology.[22] By 'phenomenology' I mean here that a phenomenon is investigated as it is experienced without convoluting it with later theoretical views that were not explicit at the historical moment. In the spirit of a phenomenology, Hegel therefore demands in the *Lectures* that each philosophy should only be described in terms of what is said in its sources (see 1986, 62). We are not allowed to ascribe positions to it that were not explicit in their historical moment. We thus limit the influence of external thoughts, and attempt to think the historical phenomenon only by itself. Each philosophy can then be reconstructed as a "stage of development" [Entwicklungsstufe] (1986, 65).

Now, Hegel pushes further into the direction of historical relativism (a term that I apply to Hegel against his proposed method): "Each philosophy is philosophy of its time. It is part of a whole chain of spiritual development" (1986, 65). This statement stands in contradiction to the idea that we can simply derive or explain history from a preconceived and universally true order of

22 To my knowledge, it is mostly Rockmore who defends this view (for example, see Rockmore 2010).

ideas. Philosophy as we know it so far is always the philosophy of its historical moment. Accordingly, we ought to derive the logic of history by internally reflecting on the ideas that were acute at their historical moment.

The phenomenological method that Hegel applies is grounded in a relativist starting point, i.e., the natural consciousness that experiences. It is, moreover, not only based on the explication of implicit conditions of a given phenomenon but also on the idea that we do not add metaphysical weight to our investigation by involving terms and concepts that are not produced on the basis of the original phenomena. So why is it controversial to understand Hegel's approach in this regard as relativist? It seems that relativism and skepticism are not sufficiently distinguished in the philosophical discourse. Hegel speaks against skepticism, while relativism was a term that was not yet used at his time. If we approach relativism as being open to the possibility of an eternal truth but acknowledging the absence of compelling arguments thus far in favor of such a truth, it becomes less controversial. Relativism, in this perspective, maintains a stance that allows for the speculative exploration of potential eternal truths while recognizing the lack of definitive evidence to support them. Given this, one might question the basis for the controversy surrounding relativism. As Rorty often indicates, there is a prevalent fear of relativism in philosophy. Beiser, for example, repeatedly emphasizes that Hegel's historicism should not be equated with relativism (see Beiser 2005, 30). In a similar vein, Pippin points out that "Hegel's outcome is not a relativist one; reasons have both a temporally local inflection and a universal aspiration" (Pippin 2008, 22). In my view, relativism is compatible with 'universal aspirations'. Philosophers are engaged in the search for truth, but their universal aspirations differ from making universal claims. In fact, Hegel's work rarely commits to explicit principles, making universal claims uncommon. Likewise, the concrete cannot be fully expressed as a universal state of rest [*Ruhe*] since it would be overly abstract. It is true that to comprehend itself, Spirit must "posit itself in externality" (Hegel 1986, 52). By situating Spirit within time and space, the historian can document the events that manifest as historical phenomena. According to Hegel's approach, the subsequent task involves stripping away the contingent aspects of the historically emerging Spirit. While this historical elimination of contingency may lead to a semblance of a universal history of World Spirit, it merely represents a new level of abstraction. Furthermore, the interpretation of World Spirit relies on the historical moment of its reconstruction since the philosophic historian is not detached from history. Being involved in history, the historian becomes a participant in the history of a Spirit that "contracts" and "expands" (1986, 56). Spirit grasps itself through new concepts provided by evolving societies. Therefore, the philosopher of history, who discovers history as a subject of interest, is part of one of the latest contractions of Spirit. They do not escape—so far—their own historicity.

Hegel, indeed, points out that it is not an aimless "infinite" cycle of expansion and contraction (Hegel 1986, 56). He even sees the progression culminating in an "imperishable essence [Wesen] of spirit" (Hegel 1986, 58). However, we have to keep in mind that the "absolute aim" of the historical "progression" (1986, 54) was defined before as the identity of spirit with itself ["das Beisichsein des Geistes"] (1986, 41). So while "the truth" that we discover is, indeed, not merely "at one time or another" (Hegel 1986, 58), this does not imply that it is independent of history. The historically developing spirit achieves its appropriate identity within a historical moment by means of the historical moment. Even though this includes eliminating time as a contingency, it remains historically generated and depends on history. The major difficulty is therefore to understand that something can be historical and timeless (this will also be the central point of my interpretation of Hegel's *Phenomenology* that is the ground for the disagreement with Habermas' interpretation).

In his *Lectures,* Hegel gives evidence for the idea of a timeless historicity. It is not, as Hegel emphasizes, that the historical forms of Spirit merely represent a hidden, deeper, and more essential content (see 1986, 54) that we stripe off like skin to reveal the truth underneath. In other words, it is not about hidden content that we have to reveal behind the curtain of history. Rather, each historical form of philosophy, as Hegel emphasizes, is itself essential (Hegel 1986, 54). History is thus not simply the surface of its underlying essence. By eliminating time, we eliminate only the contingency of an otherwise arbitrary line of events and derive the relevant history for Spirit at a certain historical moment. We then conceptually understand how history has brought us to this moment so that Spirit's aim—to understand itself—is momentarily fulfilled. It is a relative anchor, or as Siep would say, one of the slower-moving objects in the floating river of history that we, however, want to hold on to. There remains, moreover and indeed, the unlikely possibility of an end of history by this Spirit securing itself, but I do not see that this has already been convincingly achieved.

To conclude, as I understand Hegel's *Lectures,* they are an attempt to point to the sublation of the contradiction between eternal truth and historical progression. It runs as the major theme through the different versions of his *Lectures*. The solution to this contradiction, however, is not found in Hegel's ahistorical *Logic* so that an underlying set of categories can describe history independent of its historical moment. Rather, the solution is found in Hegel's *Phenomenology of Spirit* and its concept of 'Absolute Knowing', which ultimately is about a deduction of the necessary historical identity. This identity is also needed in order to start with a metaphysical *Logic*. It is about the moment of mediation, in which we finally eliminate the contingent concept of time and can understand our own involvement in history so far. My interpretation of the *Lectures on the History of Philosophy* does not rely on an interpretation of the *Logic's* 'Absolute Idea', but depends on an understanding of 'Absolute Knowing' that Hegel discusses in the *Phenomenology*. In the following parts I will therefore investigate how the *Phenomenology* prepares us for Hegel's concept of the historical identity achieved in 'Absolute Knowing' by virtue of a phenomenological and relativist historicism.

2 Hegel's *Phenomenology*

The before-mentioned epistemological problem of deriving a reliable identity is central in Hegel's *Phenomenology of Spirit*. As Rockmore, whom I largely follow in his 'non-metaphysical'[23] approach to Hegel, states: "we should comprehend his entire book as a single theory of knowledge running through different phases from cognition (*Erkennen*) to absolute knowing" (1997, 2). Hegel's epistemological account will be the ground for any further problem that we approach. Rockmore summarizes the central epistemological idea as follows: "We can never compare what is in our mind with anything outside it, but only with something else that it is given to, hence within mind" (1997, 3). While the best epistemology, in theory, would explain the universal structure of all phenomena and their relationship to mind-independent reality, Hegel holds the view that so far we can clarify truth only for us. Thus, his theory must be understood as the second best, but currently the only possible epistemology. However, although such an epistemology might be correctly considered as relativism, this does not imply that Hegel gives up on the possibility of cognition. Rockmore presents the central point of Hegel's objective cognition: "we reach knowledge when our view of the object and the object as it is given to us, or is within mind, coincide" (1997, 3). Hegel attempts to reach this identity of subject and object in Absolute Knowing. Absolute Knowing is a historical moment, in which the subject can grasp itself as its own self-made object after a series of self-investigations. The achieved identity, however, is not ahistorical.

Through the next parts of the book, I will outline the main features of Hegel's relativist epistemology leading towards Absolute Knowing. First, I will

23 Horstmann's idea is to read Hegel without the dogmatic baggage of Christian theology, which I call anti-metaphysical here.

present what I take to be key features of Hegel's phenomenology (4.2). There is a debate about the kind of account Hegel is offering. Without getting into the details of it here, I will say this: Given that Hegel chooses to focus on phenomena and to dismiss noumena, I find it adequate to call Hegel's epistemological account a phenomenological one. *Phenomenology* expresses for Hegel the idea that we deal with approachable contents of experience, and not with contents that are beyond our grasp as, for example, transcendentally deducted noumena, or mind-external things as Brandom supposes. Contents are within our phenomenal experience and we can dialectically speculate about them.

Moreover, I will also explain how Hegel's epistemology is related to subjects and their conceptual activity. Hegel's phenomenological account implies that all contents are structured conceptually. Knowledge is therefore not immediate, but mediated during the process of experience involving our conceptual activity. Hegel is interested in the possible identity of all experienced objects of this process—which, however, is not related to knowledge of a mind-independent reality. I will finish this step by highlighting how Spirit can be reconstructed as such a possible identity of subjects and their experienced objects. Spirit, moreover, represents the essence of human beings who progressively grapple with the question of knowledge, transforming this question into a matter of societies knowing themselves historically. After these preliminary interpretations, I will provide a general overview of the meaning of Absolute Spirit (4.3), before delving into a more detailed examination of the development of this concept with regard to time and historicity (4.4). Although subject and object are ontologically distinct in consciousness, they can be brought to an identity through a socio-historical learning process [*Bildung*] of spirit. Absolute Spirit, the content of Absolute Knowing, initially emerges within the intuitive practices of a community. Through progressive self-understanding, it then transcends its reliance on purely intuitive and external assumptions, ultimately conceptualizing its identity independent of time. By doing so, it attains itself as self-produced Spirit within history. It thereby delivers a criterion for knowledge relative to its particular historical moment.

2.1 *Ordinary Phenomenologies in Relation to Essentialism and Foundationalism*

As I have indicated before, Hegel's *Phenomenology* can be understood as a phenomenology, but not in a Husserlian sense. I call this *phenomenology* because it focuses on phenomena and, contrary to Brandom's interpretation, does not presuppose or deduce any noumena. More specifically, I take phenomenology to mean that *natural consciousness* is confronted with *phenomena*, while Hegel's *Phenomenology* is then the "education of consciousness itself to the standpoint of Science" (Hegel PoS §79, 50). By 'natural consciousness', I mean the subconscious production of unreflected knowledge. 'Phenomenology' refers to the method that emerges while investigating the content of *natural consciousness* and the truth of its *phenomenal knowledge* that leads to *reflection.*[24] The project is reflective since this first *natural consciousness* as consciousness investigates itself. The content of phenomenology is the description of how phenomenal knowledge is constituted through consciousness itself and will lead to the question of the consciousness experiencing itself, i.e., self-consciousness.

In order to clarify these points, I will firstly demonstrate the ordinary meaning of phenomenology as essentialist and foundationalist. After mentioning some Hegelian arguments against essentialism and foundationalism, I will discuss secondly the Kantian form of a foundationalist approach from which Hegel derives his position. I will show the beginning ideas of Hegel's phenomenology, namely how he borrows from Kant the constructivist idea of the system and how he transforms it with regard to the question of history. In the end, I will provide an argument for why content and form of Hegel's phenomenology are inseparable, and how phenomenology is a description of our conceptual, subjective activity of experiencing. Overall, Hegel's approach stands thereby in contrast with Brandom's communication model that refers to a mind-independent reality and thus exceeds the scope of what I call phenomenology.

Probably common to all forms of phenomenology is the view that knowledge emerges in the sphere of the subject, albeit under this broad definition are also gathered a lot of philosophies that are not phenomenologies (see Rockmore 2000, 51).[25] Even well-established phenomenologies are limited in their approach, as many of them consider essentialism as a prerequisite for claims to knowledge. However, Hegel challenges the notion of essentialism in

24 I call this *phenomenal knowledge*, because it is the knowledge that should be free of all presumptions about a metaphysical world, for example: that time is independent of content, hence, that time is absolute. Of course, *natural consciousness* includes contradictions since its first insights will always reveal themselves as inadequate. The following series of contradictions informs our process of education and should be understood as emerging history.

25 Rockmore, Tom. "Essentialism, Phenomenology, and Historical Cognition." *The Empirical and the Transcendental: A Fusion of Horizons.* Edited by Bina Gupta. Oxford: Rowman and Littlefield Publishers, 2000.

epistemology, arguing that it is impossible to claim knowledge without relating it to the contextual framework that justifies such claims (see Rockmore 2000, 55).[26]

Hegel's rejection of essentialism, roughly speaking, stems from the fact that language uses general [allgemeine] terms. So, we cannot simply make reference to essences that are individual. Despite the rejection of essentialism, we can still apply general concepts of language meaningfully in contrast to a social and historical background (Rockmore 2000, 55). It therefore is reasonable to suggest that we are able to operate with general concepts in sciences without any kind of essentialism.

As Hegel upholds the significance of historical contexts, he also challenges the linear foundationalism often associated with essentialism. Foundationalist approaches typically propose a secure method for establishing rigorous science after positing an Archimedean starting point of certain knowledge. This implies the paradoxical requirement of knowing what knowing is before knowing anything else. While many of these foundationalist perspectives align closely with essentialism, some approaches aim to remove the subjective perspective from current knowledge in order to uncover the conditions that make knowledge possible. Proponents of this method often believe that they can reveal the transcendental framework of knowledge or deduce the cause, the actual object, behind the appearances in cognition. They contend that we can derive such a framework by understanding how the instrument of cognition operates.

In each case, foundationalism attempts to distinguish with certainty the subjective appearance of an object from a mere semblance by setting out the rules for the 'real' science. No matter which kind of foundationalism the house of science would be built on, the foundationalist approach requires apodictic knowledge of a mind-independent object. After such a foundation, which would include a correct notion of how the instrument transforms the real object into an object of knowledge, or how essence is given directly, all epistemological problems would be solved, and from this point the real science could start (see Rockmore 2004a, 1–2).[27]

Hegel considers these approaches right at the beginning of the introduction in his *Phenomenology of Spirit*: "It is a natural assumption that in philosophy, before we start to deal with its proper subject-matter, viz. the actual cognition of what truly is, one must first of all come to an understanding about cognition,

26 Rockmore writes that it is a "mistake to think that claims to know are ever unrelated to the context in which they emerge and through which they are justified" (Rockmore 2000, 55).

27 Rockmore, Tom. *Foundationalism*. Oxford: Rowman and Littlefield Publishers, 2004a.

which is regarded either as the instrument to get hold of the Absolute, or as the medium through which one discovers it" (Hegel PoS §73, 46). We see here that Hegel distinguishes the approaches to cognition either as an instrument or as a medium. This distinction can be read as a response to the critical epistemology of Kant, even though he never mentions him by name (see Rockmore 1997, 23). Hegel acknowledges Kant's goal of developing a theory of knowledge that can ground itself, which can be roughly called 'critical'. However, he disagrees with Kant's attempt to distinguish between the content of knowledge, sensibility, and its formal apprehension through understanding. Hegel argues that the clear distinction between content and form is impossible, and thus the deduction of understanding's categories lacks sufficient means. Consequently, he concludes that the self-grounding of knowledge cannot be achieved "apart from the process of knowledge" itself, which involves consciousness experiencing (see Rockmore 1997, 24). In light of this, Hegel replaces Kant's foundationalist approach with a circular and open system, deriving the conditions of knowledge from within the process of knowledge creation. Hegel's phenomenology can thus be described as a demonstration of consciousness's experiences, as it develops in a circular manner toward self-knowledge, ultimately reaching Absolute Knowledge. In the subsequent point, I delve into the exposition of Hegel's concept of phenomenology through his critique of Kant's transcendental philosophy.

2.2 *Kant's Foundationalism and the Hegelian Rejection*

To demonstrate the differences and similarities between the phenomenological approaches of Hegel and Kant, I will explain the route from Kant's foundationalism to the roots of Hegel's phenomenology. I will sketch Kant's position and give a very brief summary of the generally acknowledged problems of his theory. I must, however, acknowledge that from the perspective of a Kant scholar, my approach might be seen as diverging from Kant's theory to some extent. Although this exploration involves a critical examination of Kant's theory, it is only done with the intention of deriving insights that can contribute to a deeper, historical understanding of Hegel's Idealist perspective. I will therefore demonstrate how Hegel derives his epistemology from a general criticism of Kant.

Kant's Foundationalism and the Idea of the System: Kant understands experience as based upon a synthetic unity of all appearances so that experiences depend on categories provided by the thinking subject. There can be no knowledge without this thinking subject to which objects appear (see Kant KdrV §16 B 141). Scientists, for example, have no direct access to mind-independent objects of nature. They interpret phenomena through an a priori categorial understanding, and therefore they depend on this categorial understanding.

By drawing our attention to the subjective conditions of knowledge, Kant makes an important contribution to epistemology. He turns the metaphysical, scientific paradigm into its opposite. Knowledge does not depend on objects anymore, but "objects depend on knowledge" (see Rockmore 2007, 60).[28] This known as Kant's Copernican revolution can also be described in the words of constructivism: knowledge is based on how we construct (see also Kant, KdrV, Bxiii).[29]

Despite this constructivism, which is anti-foundationalist, Kant goes on to make the foundationalist claim that there is a thing-in-itself outside us that affects [affiziert] our cognition.[30] The problem however is that if we can only know what we have constructed, how shall it then be possible to know that there exists a so-called thing in-itself? The commonsense reply is that there is something outside the subject causing the appearances before us. But this would mean committing a category mistake as we cannot make such a causal inference. We cannot apply a category to itself in order to ground itself. Instead, because the thing-in-itself cannot be experienced, the idea of it must be a part of the subject. However, if the thing-in-itself is related to the thinking of the subject, then it cannot be an objective condition as the older representationalist model aims for. Rather, as Kant has in mind, it must be part of a more complex holistic model which includes the subject.

For Kant, the a priori principles cannot be justified by experience, but are justified by the necessity of a whole that stands in relation to a thing-in-itself. A science of knowledge should not be a subjective conglomeration of merely related phenomena. Empirical knowledge (experience) is only possible if there is a secured unified ground available that regulates the application of our concepts. From this, his idea of a system under one unifying idea emerges. In the *Critique of Pure Reason* he writes: "In accordance with reason's legislative prescriptions, our diverse modes of knowledge must not be permitted to be a mere rhapsody, but must form a system [...] I understand the unity of

28 Rockmore, Tom. *Kant and Idealism*. New Haven: Yale University Press, 2007.

29 Kant, Immanuel. 1990. Kritik der Reinen Vernunft. Hamburg: Felix Meiner.

30 I recognize that my criticism of Kant is somewhat simplified, particularly considering that the thing-in-itself can be interpreted as a regulative idea or a limit idea. Nevertheless, it is crucial to highlight that Kantian theory has given rise to conflicting interpretations, encompassing both foundationalist and anti-foundationalist perspectives, as expounded in Rockmore's book *On Foundationalism* (2004). Hence, my intention is not to present the most accurate account of Kant's philosophy but to outline the conflicts that have motivated Hegel's theory.

the manifold modes of knowledge under one idea" (Kant CpR B860, 653).[31] In other words, Kant wants to unify all diverse modes of our knowledge under one idea. This idea is the rational concept as a whole, according to which all concepts are organized. But since this rational concept must be provided by the subject, the question of the unity of the subject becomes central to his investigation. Kant relies on the formal I-think, which must be able to accompany all representations and which is supposed to support the coherence of all representations with respect to a possible mind-independent reality.

In my view, Kant does not solve the question of whether objective claims with respect to a mind-independent reality can be made through finite, human beings. He presupposes a formal subject that projects a mind-independent object as the absolute condition for knowledge. The epistemological subject becomes central to his approach, while the experiencing subject receives a subordinated position in his investigation. This is because the experiencing subject can only know phenomena, whereas the epistemological subject can think the noumenon as a necessary transcendental point of convergence giving unity to our cognition. Kant's theory is a priori, hence ahistorical.[32]

Kant wants to leave the circle of the experiences of the subject because experience is subjective. But this entails a formalization of an epistemological subject that is different from experiencing human beings. Regarding the subject of cognition, he thus defends a dualism between the mechanisms of cognition. Sensibility is the passive medium, and understanding as the instrument organizes these phenomena through a priori categories. This diversity of the subject, however, is the problem for the unity of reason. In order to overcome this dualism of the two stems of cognition, Kant attempts to combine both sides in one system. He concludes that sensibility and understanding must be part of a system of knowledge in order to make it coherent. This means, on the one hand, Kant regards the sensory manifold as the medium in which subjects receive information; on the other hand, he sees understanding as the instrument through which the information is spontaneously categorized and through which an object of experience is brought up (see Rockmore 2005, 24). The general idea of this approach is to understand how cognition in identity works. In order to justify this system, Kant believes that he can subtract the subjective perspective from appearances in order to deduce the conditions of

31 Kant, Immanuel. *Critique of Pure Reason.* Translated by Norman Kemp Smith. Palgrave Macmillan, 2007.

32 Kant's philosophy is supposed to be the first and the last philosophy. For instance, Kant states that any change in his philosophy "introduces contradictions not merely into the system, but into universal human reason" (Kant CpR, Bxxxviii).

the possibility of knowledge, the a priori categorial framework. The categorial framework, which is necessary for experiences, is only possible if it produces the idea of a real thinkable object, the thing-in-itself. Kant therefore follows a foundationalist goal, namely to justify a representational system under one unifying, objective, and a priori idea. This goal, as explained above, presupposes a separation of content and form, while we focus on the formal aspects that are necessary to order content.

So far, I have only sketched how the Kantian system relies only on an epistemological subject. In the following points, I would like to show how Hegel evaluates and handles Kant's framework, which in my interpretation is the root of Hegel's phenomenology. Contrary to Kant, Hegel argues that content and form are not clearly separable.[33] Moreover, he abandons the idea of a formal I-think and a noumenon. Nevertheless, Hegel borrows Kant's constructivist idea, and the idea of a systematic structure of our knowledge. In this sense, Hegel is a post-Kantian thinker who proceeds with the anti-foundational project of Kant by reinterpreting the Copernican turn with regard to the question of history.

Hegel's Interpretation and Rejection of Kantian Cognition: Hegel comments on Kant's idea of separating content and form:

> For, if cognition is the instrument for getting hold of absolute being, it is obvious that the use of an instrument on a thing certainly does not let it be what it is for itself, but rather sets out to reshape and alter it. If, on the other hand, cognition is not an instrument of our activity but a more or less passive medium through which the light of truth reaches us, then again we do not receive the truth as it is in itself, but only as it exists through and in this medium.
>
> HEGEL PoS §73, 46

As Hegel states, it is obvious that an instrument alters the object, and also the medium shapes the way we perceive it. It seems, nevertheless, that we could deduce the real object through correct knowledge of the instrument and the medium. But Hegel discards this idea:

> If we remove from a reshaped thing what the instrument has done to it, then the thing—here the Absolute—becomes for us exactly what it was before this [accordingly] superfluous effort.
>
> HEGEL PoS §73, 46–47

33 Hegel's theory is therefore incompatible with Brandom's account which develops a formalized language game that presupposes a mind-independent reality.

So, if we subtract what the instrument has altered, then we just receive what we had before, an unknown object. Our effort was useless. Moreover, to presuppose that cognition as the instrument is bringing the absolute closer to us presupposes that there is "a merely immediate and therefore effortless relationship" to the absolute (Hegel PoS §73, 47). The question would then be: why did our cognition alter it in the first place?

The same is true for the approach to apply the law of refraction to the concept of cognition as a medium (see Rockmore 1997, 25). When we "determine" the "deviation of this medium", and we subtract this deviation from the object which is in the medium, then we know something about the process of cognition, but not about the object in-itself (see Rockmore 1997, 25). Hegel writes: "For it is not the refraction of the ray, but the ray itself whereby truth reaches us, that is cognition; and if this were removed, all that would be indicated would be a pure direction or a blank space" (Hegel PoS §73, 47). In other words, understanding the medium's refraction would just produce a description of the way the object reaches us and bring us back to where we were before the object reached us. By this, we would not gain any more information about the object (see Rockmore 1997, 25).

The arguments presented suggest that content and form are inseparable; it is impossible so far to establish a categorical a priori framework before experiencing content. Hegel's view is that cognition operates effectively, leaving no need for doubts about how consciousness experiences. He places trust in our ordinary way of being in the world, and there is no necessity for a pre-established method to embark on the path to science. Knowledge begins with experience, but it transcends mere empirical encounters. Hegel's project involves developing an approach to derive the correct conclusions from experience, thus encompassing a broader understanding of knowledge.

Kant's distrust in cognition is, in Hegelian terms, "the fear of falling into error" (Hegel PoS §74, 47). For Hegel, the result of the Kantian theory would be skepticism. Hegel specifies his rejection of Kant's account: "it takes for granted certain ideas about cognition as an *instrument* and as a *medium,* and assumes that there is a *difference between ourselves and this cognition*" (Hegel PoS §74, 47). Hegel rejects this dualism of cognition as it divides the experiencing subject and the epistemological subject. This leads us to a formalized transcendental subject as well as a transcendental object. Furthermore, Hegel writes about such an artificial separation: "Above all, it presupposes that the Absolute stands on one side and cognition on the other, independent and separated from it" (Hegel PoS §74, 47). In other words, Hegel rejects Kant's method because it creates an ontological gap between the Absolute and cognition.

Nevertheless, Hegel does not give up entirely on the Kantian 'system'. Although he demands that we should "mistrust this very mistrust" of Kant in our experience (Hegel PoS §74, 47), Hegel focuses on the conditions of knowing that are within the experiences of finite, human beings. For Hegel, human beings already have an understanding of the world, even when they are not grasping a mind-independent reality. Rather than searching for the a priori conditions of science, it is reasonable to accept that human beings were always capable of dealing with objects. Therefore, Hegel states:

> One may set this aside on the grounds that there is a type of cognition which, though it does not cognize the Absolute as Science aims to, is still true, and that cognition in general, though it be incapable of grasping the Absolute, is still capable of grasping other kinds of truth.
>
> HEGEL PoS §75, 47

In other words, Hegel looks closer at the absolute which is related to our being in the world and as a consequence, the Hegelian project has to be described as a theory a posteriori. It observes human beings as within the absolute and not separated from it. Since we are already in the truth through our being in the world, our task is to make explicit the concepts of our natural consciousness that is in a certain way already related to the absolute.

The method for this task would be phenomenology, or to put it in the proper terms, to make explicit what is phenomenal knowledge of the experiences of consciousness (see Marx 1975, 7).[34] This form of investigation, however, can only be understood as related to the content of the investigation, natural consciousness.

Phenomenal Knowledge of Natural Consciousness and the Concept [Begriff]: I have sketched how Hegel rejects the Kantian foundationalist project which contains many disputable presumptions such as the thing-in-itself as the cause of appearances, dualistic views about cognition as an instrument or as a medium, or the ahistorical a priori categorial framework that is unified through the formal I-think. Instead, Hegel begins with an investigation of natural consciousness (Rockmore 1997, 26). Here the concept of Hegel's phenomenology becomes concrete: we observe the phenomenon of experience. If we always

34 Marx, Werner: *Hegel's Phenomenology of Spirit.* Translated by Peter Heath. Chicago: University of Chicago Press 1975.

possess a certain kind of truth in our natural consciousness, then the question arises, what we have to make explicit? Hegel thinks that we have to make the Concept [der Begriff] explicit.[35] He writes: "For to give the impression that their meaning is generally well known, or that their Notion [Begriff] is comprehended, looks more like an attempt to avoid the main problem, which is precisely to provide this Notion" (Hegel PoS §76, 48). As we can see, the basic epistemological task is to explicate the Concept [*der Begriff*]. Hegel discards the presupposition of a noumenon and the 'I-think', replacing them with the idea of the Concept, which must be historically attained. In contrast to Kant and his notion of unifying the given, Hegel perceives the unifying activity as nothing more than natural consciousness. His strategy is to elevate natural consciousness to the level of the Concept, which represents the absolute and by which we improve our grasp of this unifying activity (see Marx 1975, XX). Natural consciousness is therefore the ground for the unity of the object for us. This ground is constituted through the way in which consciousness works, and therefore we are not talking about a linear foundationalism which tries to prove the existence of a real object and secures in this sense a science of mind-external reality; rather, we are talking about a circular system that has its starting point in natural consciousness and its experiences.[36] We thus enter an anthropological investigation of what humans are and how they constitute knowledge for themselves.

Consciousness understood through the Concept [der Begriff], moreover, is not limited to a concrete, conceptual shape. It is a major and often-committed mistake to attribute the position to Hegel that humans 'are' concepts. Rather, concepts are the tool that we have available for our project of knowledge. To understand subjects by virtue of their conceptualizing activity has some advantages. Concepts symbolize the reflective relationship that natural consciousness builds to contents. By assuming that concepts are necessary tools for

35 Hegel's 'der Begriff' will be translated as the Concept in the following. The translation that I use says 'Notion,' which trivializes its meaning.

36 I adopt Rockmore's proposition that Hegel's established system is circular, which naturally leads to the question of how progress is possible in the history of consciousness. However, it is essential to recognize that the closure of the circle elevates the circular element to a higher level, negating the former viewpoint. We can call this progress, even though, according to Hegel's contentious concept of sublation [*Aufhebung*], this process also preserves the previous viewpoint in a transformed way. I acknowledge that this presents a contradiction, which can be approached in two ways: one can either attempt to resolve it by introducing a new concept, or, in light of recent developments in paraconsistent logic, justify it as a single contradiction being true without causing the entire system to collapse due to just one or more limited exceptions.

grasping anything, for example, we already admit that there is not an immediate relationship to a mind-independent object. Concepts express the fact that we have a mediated relationship to content. Concepts are therefore also not empty, mere universals. They always presuppose content. Understanding this relationship might clarify what can count as knowledge for us. Acknowledging this, however, we have to break with the idea of recognizing a mind-independent reality because finally each cognition that emerges in the subject is "permeated through the categories of the self" (Marx 1975, XXI). Due to this fact, Hegel will undertake "an exposition of how knowledge makes its appearance" (Hegel PoS §76, 49). In this process, natural consciousness takes itself as real knowledge, but has to discover that on "the pathway of [...] despair" nothing is left but phenomenal knowledge (Hegel PoS §78,49). Thus, it reaches the insight that all reality is in "truth only the unrealized notion [Begriff]" (Hegel PoS §78, 50). For the epistemological project, this means: there is nothing to find but language that has developed historically. Basically, Hegel's theory means that we have to understand how we can use concepts for our project of knowledge.

Though this focus on concepts might resemble the linguistic turn, Hegel goes beyond the linguistic and pragmatic turn by including the concept of historical spirit. He describes language as a particular human language, resulting from human interactions in a historical community. So, Hegel does not deliver a formalized model of communication, but he investigates the dependence of content and form in relation to how human beings communicate historically. This critical knowledge of our language can be reached only through a series of stages which Hegel signifies as the "detailed history of education" (Hegel PoS §78, 50). It is the history of human kind reaching the stage of science (see Rockmore 1997, 27). In this history, rejections of theories are just the consequence of the progression within the series of these stages. For instance, every negation of a predecessor has to be understood as a result of the tradition, and Hegel calls this concept "determinate nothingness" (Hegel PoS §79, 51). Therefore, the progression, to put it simply, happens relative to history. The final stage of this investigation would be the concept of the Absolute, which is the Concept of the Concept. Hegel's phenomenology is then to show how consciousness works conceptually, how consciousness relates itself to phenomenal knowledge [erscheinendes Wissen] as real knowledge (how the thing is for itself) by virtue of the Concept, and how it, at the same time, distinguishes itself from this knowledge, because it has the insight of the phenomenal appearance of things. All of this will be demonstrated as a function of the Concept that can relate to itself in a historical process.

Under these premises, reality is still a concept that is vital to our processes of knowledge. Yet, it is not understood as a mind-independent reality, but as a reality that is produced within the movement of the Concept. Since this movement is a movement of concepts, Hegel's project can be, indeed, called a linguistic project. This, however, does not mean that it is exhausted by the insights of the linguistic-pragmatic turn. It is not a linguistic analysis of contingent concepts, but it is also about the historical appearance of linguistics as the current core of our research. Hegelian philosophy expresses its own historical emergence through the Concept of the Concept. Often taken as the final form of his theory, it is, however, within the boundaries of a particular, historical moment. The Concept of a Concept means then that it cannot be superseded by the means of the particular historical moment that, in this instance, is concerned with language. In this sense, the linguistic turn is anticipated in Hegelian philosophy. It does not imply, however, a secured epistemology a priori. It neither expresses that our worldview is ultimately limited to one universal form of this conceptual activity for all history. Language and the Concept of the Concept are rather the highest mediations with regard to the historical epistemological problem that was known at Hegel's historical moment and possibly to our historical moment. It is a historical standpoint of science. While we could say that current epistemology is brought to an end by it, in the sense that we cannot go further beyond, it still delivers only the preliminary, epistemological grounding for our current ontology that may develop beyond.

2.3 *Hegel's Account of Self-Consciousness and His Epistemology*

In the last point, I highlighted Hegel's relation to his predecessor Kant. I have presented his view that content is not independent of form. Hence, we cannot simply approach mind-independent objects. Instead, our epistemology is related to the mind-dependent reality of experiencing human beings and their history.

In the following point, I further discuss Hegel's introduction to the *Phenomenology of Spirit.* I demonstrate how his epistemology is rooted in the investigation of consciousness. Before Hegel, with the possible exception of Fichte, the subject was only understood with regard to the requirements of knowledge but never discussed as a full person (see Rockmore 1997, 4). After Hegel, the subject becomes detranscendentalized and is investigated as an experiencing human being in history. The historical epistemology that is developed on this ground will culminate in the exploration of Absolute Knowing, a subject that will be discussed in the subsequent part (4.4) of this chapter.

Hegel, acknowledging Descartes for his transition into subject philosophy, criticizes him for only employing "canons of evidence that were uncritical and untested" (Greene 1972, 19).[37] For Hegel, the problem is that knowing is already altered by consciousness and thus certainty cannot be provided by the subject itself (see Greene 1972, 19). Instead, consciousness needs to be recognized in its acts, and not in its persistence through time. Kant, succeeding Descartes, already "enunciated the supremely important principle of the identity of ego and Notion: the pure forms of thought, i.e., the forms of the Notion" (Greene 1972, 21). For Kant, concepts are forms of the *self-producing*, unifying activity of the thinking consciousness (see Greene 1972, 21). Kant, having made this shift to the conceptual understanding of cognition, however, did not deliver "a genuine demonstration of his great principle" (Greene 1972, 21). The question remained of whether we can move beyond an "abstract, empty idealism" and build a relationship between a formalized unity of logic, and its contingent, and empirical content, the 'Realphilosophien' (Greene 1972, 21). Historically, Fichte's abstract idealism could also not exhaustively explain the relationship between the 'I' and the contingent content. In Fichte's theory, content remained as an external impetus [Anstoß]. Thus, we stayed within the boundaries of that unexplained dualism.

For Hegel, this contradiction of a rational self and an irrational outside—for example, the assumption of an in principle unknowable noumenon—is itself irrational (see Greene 1972, 21). As a consequence, Hegel does not start with a transcendental subject. He rejects the idea of a pure, self-evident ego that remains always the same and only confronts itself with irrational content (see White 1983, 24). The question, however, remains: how can we achieve unity between the content (also called the truth, or the object), and our knowing (also called our activity, or the subject)? Hegel's solution is that the unification is found in human beings and their ways of knowing themselves as humans in a historically developing society. For this reason, Greene writes: "The *Sache selbst*

---

37 Like any thinker, Hegel's epistemological project can also be accessed in relation to its predecessors. I regard the most important task for Hegel to overcome the Fichtean opposition of content and form in consciousness and achieve an absolute mediation of the further occurring contradictions that such a project reveals (see Greene, Murray. *Hegel on the Soul—A speculative ontology.* Hague: Martinus Nijhoff, 1972, 24). The last mediation in this process will be man himself who is exposed to the command "know thyself" in an epistemological sense. This means that Hegel does not believe in an isolated, pure ego like Fichte (see White, Alan. *Absolute Knowledge: Hegel and the Problem of Metaphysics.* Athens: Ohio University Press, 1983, 24). Thus, his beginning is the incomplete, historical status of man who faces the problem of knowledge that is supposed to be universal. This subjective condition, its historical starting point, must therefore be included in a successful project of thought (see White 1983, 125), and its consequences for historicism have to be indicated (see White, Alan. *Absolute Knowledge:Hegel and the Problem of Metaphysics.* Athens: Ohio University Press, 1983).

for our particular study is man himself, who is commanded to know himself by the law of his own being as Spirit" (Greene 1972, 21). In other words, Hegel's epistemology is based on a subject understood anthropologically. I will explore this changed view of the subject and its epistemological role with regard to the introduction of the *Phenomenology of Spirit* in the following paragraphs.

Hegel critiques the inherent incompleteness of the project of science, asserting that "Science, just because it comes on the scene, is itself appearance: in coming on the scene it is not yet Science in its developed and unfolded truth" (Hegel PoS §76, 48). Recognizing this incompleteness, Hegel argues that our starting point should be natural consciousness, which is engaged in the process of knowing but has not yet reached the standpoint of science as its ultimate goal (see Rockmore 1997, 26). Thus far, we have been unsuccessful in elucidating the conditions of knowledge. Furthermore, Hegel raises doubt as to whether natural consciousness can truly attain scientific knowledge. However, he clarifies that this doubt does not lead to skepticism (Hegel PoS §79, 51). In one of the many dialectical twists within the book, Hegel observes that doubt in natural knowledge is always a doubt of something, which ultimately yields a positive outcome. Hegel describes this positive outcome as "determinate nothingness" (Hegel PoS §79, 51). The term signifies the result of an ongoing and self-perpetuating process of scientific inquiry through negation. In this self-developing process, previous forms of knowledge are negated to make way for new ones. Ultimately, this process of negation will lead to the conception of philosophy as the highest form of cognition within our known history.

Hegel emphasizes that this movement has a goal, stating: "But the goal is as necessarily fixed for knowledge as the serial progression; it is the point where knowledge no longer needs to go beyond itself, where knowledge finds itself, where Notion corresponds to object and object to Notion" (Hegel PoS §80, 51). The goal of the process of knowledge is the fulfillment of an identity criterion, specifically, the completeness of the Concept. Rockmore sheds light on this aspect, highlighting that Hegel's approach relies on the subject's ability to differentiate between its subjective view of the object and what is given in conscious experience, presupposing a conception of self-consciousness (1997, 29). To reiterate, the goal of the process of knowledge is the fulfillment of an identity criterion, i.e., completeness of the Concept. But why is the goal the concept? Because the concept can imitate the complex dynamics of consciousness, self-consciousness, and its self-reflective relationship with content, Rockmore elaborates on this aspect of the subject's reflective capabilities:

> Hegel's approach rests on the subject's ability to distinguish between its view of the object, roughly what the subject thinks the object is, or its

> 'theory' about it, and what is given in conscious experience. This conception presupposes a conception of self-consciousness.
>
> ROCKMORE 1997, 29

This citation requires a longer explanation. Naturally, the standard criterion for the identity of the concept and its object is the object in-itself (see PoS §81, 52). However, the object in-itself is not directly accessible. Hegel claims that we will see how knowledge, the object for us, and the object in-itself, which Hegel takes to be truth (see PoS §82, 52), occur in consciousness as necessary moments. We can compare both of these moments, since the ontological difference is bridged by consciousness itself. The object will be revealed as an externalization of consciousness so that consciousness grasps two objects: an object experienced by consciousness and an object idealized or reconstructed by consciousness. The important point is that both of these are linked to consciousness. In order to achieve the standpoint of science, we therefore have to grasp consciousness, which means developing a conception of self-consciousness. Self-consciousness involves consciousness reflecting upon itself and discerning its own distinctness. Hegel explains this self-distinction within consciousness in this manner:

> Consciousness simultaneously *distinguishes* itself from something, and at the same time *relates* itself to it, or, as it is said, this something exists *for* consciousness; and the determinate aspect of this *relating*, or of the *being* of something for a consciousness, is *knowing*. But we distinguish this being-for-another from *being-in-itself*; whatever is related to knowledge or knowing is also distinguished from it, and posited as existing outside of this relationship; this *being-in-itself* is called *truth*.
>
> HEGEL PoS §82, 52–53

So, consciousness by a constant process of negation is always reaching beyond itself. Ultimately, it is a process of self-negation, and thus, it does not retain what it has but strives for the standard of an in-itself to extend its own limitation. It is consciousness that posits a distance to the object and it does so by self-negation. Thereby, it projects objects in themselves and thus relates to them by virtue of projection. However, in Hegel's framework, the object in-itself differs from Kant's noumenal, mind-external object. It refers to the idealized object that consciousness synthesizes based on previous experiences. In other words, the object is derived from past experiences of consciousness. As consciousness constantly emerges from the past, it carries synthesized memories of objects and compares them to the present moment. In the following citation, Hegel discusses the internally emerging criterion that consciousness produces in this process: "Consciousness provides its own criterion from within itself so that the investigation becomes a comparison of consciousness with itself; for the distinction made above falls within it" (Hegel PoS §84, 53). This marks a significant shift in the perspective on criteria within the history of philosophy. Hegel reveals that the criteria for knowledge are provided by consciousness itself. Notably, there is no reliance on an external criterion. In this process, consciousness produces its own epistemological criterion, distinguishing itself as both a past consciousness and a present consciousness. The goal to Hegel is to cease the continuous movement of consciousness by comprehending how consciousness generates its object.

A point has to be made on the project of Husserl's or Heidegger's phenomenology that I have to distinguish from what I see as Hegel's phenomenology. It seems that in order to understand the described experiential process of consciousness, we require a phenomenological reduction so that the object can appear as it is. This phenomenological reduction, however, would presuppose a mind-independent object that is supposedly altered by the misguided attitudes of an involved consciousness, or that is distorted by existential categories. Against these particular phenomenological programs, I have to say something about their presupposition that things can show themselves. Presupposing self-exposition would be an anthropomorphism that attributes human activities to things without considering our subjective activity as a part of the constitution of objects. Can things show themselves? Perhaps, but if things do not show themselves, a phenomenological reduction is misleading. Hegel avoids this problem and focuses instead on the object that is already in consciousness and externalized by consciousness. For Hegel, the in-itself object is a presupposed relation of consciousness to the essence of an object, while the concept of it is knowledge. Knowledge is true if essence and concept are equivalent to each other. Since the truth—the mind-internal, experienced object—can, however, change, the criterion for their correspondence can change as well. Thus, Hegel proposes a fundamentally different phenomenology. We are not getting closer to reality or essences, but we act coherently with regard to the criterion that our consciousness has achieved: a concept corresponding to the essence of a mind-internal object that ultimately is the activity of consciousness itself. Truth is then rather a question of coherently acting with regard to our experiences that are within consciousness. In this way, it makes sense to say we get closer to truth *for us*, while mind-external truth is not the target of Hegel's discussion. Moreover, Hegel's view on the in-itself and the for-itself is completely different from Brandom's view who believes that the in-itself stands for mind-external reality.

Beyond Hegel revising the traditional view on truth, Hegel also reflects on the idea that objects depend on the theories we hold about them. Hegel emphasizes this phenomenon with the following statement: "If the comparison shows that these two moments do not correspond to one another, it would seem that consciousness must alter its knowledge to make it conform to the object" (Hegel PoS §85, 54). In the following paragraphs, Hegel suggests that not only do theories change, but the objects themselves undergo a transformation. Brandom struggled to reconcile this aspect and argued that it would portray Hegel as an "implausible Idealist" if taken literally (2019, 90). Contrary to Brandom, I believe, it encompasses a theory that underscores the intrinsic interconnection between the nature of objects and our theories. With regard to experience, Hegel explains the alteration of objects:

> This is the moment of transition from the first object and the knowledge of it, to the other object, which experience is said to be about [...] the new object shows itself to have come about through a *reversal of consciousness itself.*
>
> HEGEL PoS §87, 56

So experience is the process in which the object changes. Hegel does not expound on how this process of experience works in detail (see Rockmore 1997, 33). I assume that since the criteria for knowledge can be different (see Hegel PoS §86, 55), the process of experience might be different for particular objects. Moreover, Hegel does not say that experience is about a mind-independent reality, but he talks about how objects can be constituted only with regard to our experience which is—to some degree—mind-dependent. It is within this context that I interpret Hegel's philosophy through the notion of 'empirical realism'.

With regard to empirical realism, Rockmore distinguishes three kinds: first, "direct knowledge of an independent object", second, "indirect knowledge of an independent object through a dependent object", and third, the denial of direct and indirect knowledge claims of a mind-independent world, but experience of mind-dependent objects (see Rockmore 1997, 197). Given the former arguments, Hegel commits strongly to "tertiary empiricism" (see Rockmore 1997, 197). Experience is then the process of the emergence and production of new objects. In the previous points, I showed Hegel arguing consciousness is only consciousness of its object and that form does not exist independent of content. From these two premises and this particular understanding of empiricism, it follows that consciousness must be altered in the process of experience. Since, however, mind-internal objects do not exist independently of consciousness a change in consciousness must also change the object. This implies that the dialectic of consciousness cannot be brought to a halt. It starts with experience, but it initiates a cascade of further changes in our view of objects. Our changed views, however, will eventually be challenged by new emerging objects which constitute new experiences. In conclusion, we see Hegel improving our understanding of cognition by adding to it a notion of development. Reflecting on the *history* of one's experience and coming to know things in relation to this history we come to a better understanding of what qualifies as knowledge for us.

With regard to our construction of empirically based sciences, Hegel makes clear that we do not simply change our theory if a new object arises. Instead, the new object as it will be reconstructed contains its predecessor in the negation. By virtue of this positive negation, Hegel grasps the "nothing *of that from which it results*" (PoS §87, 56). It is a result that "contains what was true in the preceding knowledge" (PoS §87, 56). This is incompatible with an approach of falsification. According to the approach of falsification, we assume that a change in the perceived object ultimately denies our theory so that we would simply replace the former theory. Yet, we do not simply replace the former theory. Instead, Hegel argues that we ought to understand our theories as developing historically. They are an integral part of what makes the objects that we have at hand. What is under discussion is not mind-independent reality, but rather what human cognition is in its historical development. In this sense, we could say that the transcendental and empirical consciousness of Kant come together in one theory—a theory of an ever evolving consciousness based on its history of experiences that is, at the same time, the ground for this history of changes.

We can already see how the concept of historicity is contained in the simplest acts of knowledge. This means that each cognition stems from a process of comparing the in-itself object with the for-itself-object, while each of these objects are the results of former comparisons. Change, moreover, perpetuates further changes. In other words, there is a history of comparisons between the object given in the present consciousness and the theories about the object. Theories change if they are shown to fail and become more adequate theories about the contents of experience.

As consciousness continually alters both its theories and objects, it becomes questionable whether it can ever find complete satisfaction. However, Hegel posits the possibility that consciousness may acquire knowledge of its own knowing. Yet, in the search for an inner resting point, consciousness encounters a conundrum: on one hand, there is a consciousness that projects previously experienced objects as 'objects-in-themselves'; on the other hand, there is a consciousness that experiences present objects that diverge from the projected ones. These two consciousnesses serve different functions, and the goal is to explain them as integral parts of a unified consciousness.

As Hegel does not believe in ahistorical accounts, his theory of self-consciousness cannot, for example, be a Kantian transcendental and necessary point of convergence. His theory does not conclude with the notion of an eternal scientific self-consciousness existing independently of further action. Rather, his theory of self-conscious subjects is derived from individuals situated within their historical moment, who actualize their potential and shape their lives as expressions of a historical Spirit. Their self-realization resulting in Absolute Knowing—the knowledge of this historical Spirit—is arguably the most contentious aspect of Hegel's theory. Particularly, Habermas identifies here Hegel's ultimately ahistorical account. The question arises, indeed, as to whether Absolute Knowing represents an ahistorical endpoint in Hegel's system or if Absolute Knowing represents another moment within history. Confusion may arise when considering the term 'Absolute' used to describe this form of knowledge. In my interpretation, it is absolute, but only in the sense that it cannot be surpassed by current means within our historical moment. Thus, it serves as a provisional criterion proposed for our present historical context. It is not a formal and definitive criterion and remains open to further transformation contingent upon the historical experiences of human beings. As a reflection on our current history, however, and its inclusion of everything that can be experienced, it is the highest attainable stage in our society at this historical moment. In the following parts, I reconstruct Hegel's concept of Absolute Knowing and demonstrate how my interpretation of this concept circumvents the epistemological predicaments of mind-external realism and ahistoricity.

3 Absolute Knowing in Relation to History and Time

To distinguish Hegel's Absolute Knowing from Brandom's and Habermas' ahistoric accounts, we must examine Hegel's differentiation between time and history. Time is primarily a cognitive tool for organizing events on a timeline. While historical events occur in time, their essence goes beyond this linear progression. Our grasp of history depends on discerning meaningful conceptual connections among pivotal historical events. For instance, constructing a causal sequence of historically meaningful events necessitates concepts that transcend a mere linear progression. In essence, history is best comprehended through the intricate web of conceptual associations among a selection of significant events. Mere chronology provides no insights into their interconnectedness.

In this chapter, I aim to show that while time helps organize empirical phenomena, historicity lets us understand the essence of their historical becoming. In this regard, Absolute Knowing represents the highest stage of the reflection on our own history, marking the end of a historical era without, however, bringing history itself to an end.

At this point, it is crucial to clarify that the specific aspect of Spirit reaching Absolute Knowing differs significantly from the transcendental subject envisaged by Hegel's predecessor, Kant. Hegel does not assume the existence of a transcendent soul. He also does not assume a spirit orchestrating the world's unity for us. Instead, Spirit is discernible as a force within world history, abstracted from a complex dialectical interplay between experiential, incomplete consciousnesses and their pursuit of universality. This relationship, characterized by its incompleteness and aspiration towards truth, undergoes a historical evolution in the context of world history. Hegel delineates the culmination of this particular trajectory in the closing sentence of the introduction to his *Phenomenology*: "finally, when consciousness itself grasps this, its own essence, it will signify the nature of absolute knowledge itself" (Hegel PoS §89, 57). This specific self-consciousness unfolds as Spirit progresses through various stages within historical communities (see Westphal 1979, 218).[38] In the end, it attains self-identity when it recognizes itself as a historical Spirit actively shaping itself within the confines of its historical moment. Consequently, my thesis posits that Absolute Knowing provides insight into ourselves as beings who actively shape history within the boundaries of their historical contexts. Through this process, they establish standards for their own development. I will initially explore this interpretation in the context of distinguishing between world history and its relationship with Absolute Knowing, particularly with reference to Jaeschke's interpretation.

3.1 *Hegel as a Thinker of World History and Absolute Knowing*

One of the major problems is at this point to explain how Absolute Spirit emerges within and through the history of our world, i.e., world history. On the one hand, Absolute Spirit can be seen as the content and eternal substance of the course of history, while world history is its form. On the other hand, Absolute Spirit must be a part of world history because Hegel, as we have

38 Westphal, Merold. *History and Truth in Hegel's Phenomenology*. Bloomington: Indiana University Press, (1979) 1998.

discussed before, rejects the oversimplifying distinction of content and form. This means that Absolute Spirit is the objective essence of world history, while, at the same time, it is also its medium of emergence, i.e., world history.

This tension is also related to the difficulties of historicism. In order to be relevant as a theory that can deliver knowledge, historicism needs to establish a criterion of objectivity. However, as long as we cannot claim a standpoint outside of history, this criterion of objectivity has to be developed within a historical framework. Rockmore gives a clear explanation of this issue:

> [Historical relativism,] which refers to our incapacity to escape from the limits of our own historical moment, can be taken as suggesting that claims to know are objective but also historically relative. A knowledge claim that was historically relative but not objective would be uninteresting; and a knowledge claim that was objective but transhistorical would not be historically relative.
>
> ROCKMORE 2005, 195

This implies that Hegel qualifies only as a historical relativist if all his claims are relative to the historical moment. If Absolute Knowing is independent of its historical moment, then Hegel's theory could not be identified as such a kind of historicist relativism. If his historical relativism, however, excluded objectivity, his theory would fall back into skepticism. With regard to Absolute Knowing, it comes down to the question of how we can achieve communal, objective cognition in history, and how the achievement of this cognition must be understood with regard to the limits of our specific *world history*. According to my relativist interpretation, it is the question of how Absolute Spirit can achieve objective cognition and, at the same time, be bound to its historical frame.

Once again, I would like to emphasize that the distinction between time and history plays the crucial role here because it will lead us either to a relativist or anti-relativist position. Spirit emerges through time which is a succession of temporal stages. If we understood the history of Spirit as the emergence in a merely temporal framework, its events would only be contingent points on a timeline. However, history is the more fundamental concept. According to the concept of history we organize and conceptualize the stages that occur in time. The distinction between time and history eventually occurs in Hegel's chapter on Absolute Knowing. According to the general argument there, we come to an understanding of history by eliminating external elements that inhibit a superior, conceptual understanding. Time is accordingly grasped as the representational, intuitive understanding [Vorstellung][39] of Spirit that eventually has to be eliminated. The subsequent controversy revolves around Hegel's goal to eliminate, or in his words, 'to annul' time. As we have seen, Habermas thinks that Hegel's concept of Absolute Knowing does not annul only time, but also annuls history. He therefore criticizes Hegel's project as positing itself as eternally true, i.e., as anti-relativist. My reading of the passages will reveal, however, that Hegel defends a particular eternity that is self-produced by subjects in a community. This self-production is absolute only in the sense that it does not require any external foundation. It is given by humans to humans through their experienced history with their social and historical reflection on it—we can call this their historicity. Hence, Hegel does not return to a transcendental realism, as Habermas assumes, but rather turns to a concept of reality that is the result of a communal, historically developing, subjective activity. The subject remains embedded in its historically, experienced reality, while time as an intuitive form is eliminated only from its conceptual understanding. In fact, we intend to arrive at a true conception of a historically reflecting subject, which is at the same time, the historical object and origin of this investigation, i.e., historicity.

Habermas' interpretation of Absolute Knowing as an ahistorical subjectivity is not an uncommon perspective. Hoffmeyer, among others, highlights that many scholars have observed Hegel's apparent shift towards ahistoricism, especially in the final chapter of the *Phenomenology*:

> According to the Phenomenology of Spirit, Spirit becomes itself only in and through history. Yet the concluding section on 'Absolute Knowing' has left many interpreters wondering whether Spirit in Hegel's eyes is only penultimately historical. Does Hegel ultimately relativize Spirit's historicity within an ahistorical framework?.
>
> HOFFMEYER 1992, 198[40]

Fulda also pays special attention to the last chapter and focuses on the tension between Absolute Spirit and its historical becoming. According to him, many critics (particularly Habermas) believe that Hegel's project failed because of the final chapter of his *Phenomenology* (see Fulda 2007, 340).[41] Fulda rejects

39 Also translated as 'picture-thinking'.

40 Hoffmeyer, John F. "Absolute Knowing and the Historicity of Spirit." *The Journal of Religion* 72 (2), 1992: 198–209. http://www.jstor.org/stable/1205149.

41 Fulda, Hans Friedrich. "Das absolute Wissen—sein Begriff, Erscheinen und Wirklichwerden." *Revue de Métaphysique et de Morale* 55 (3), 2007: 338–401. DOI 10.3917/rmm.073.0338.

Habermas' criticism. He believes that the last chapter of Absolute Knowing needs to be reconstructed with a focus on its openness towards experience and historical change.[42] In the spirit of Fulda, I would like to interpret Hegel's concept of history with regard to its openness. To begin, I will offer a general interpretation of what I perceive to be his historicism, later (part 4.4) I will fill in the details with regard to a concrete interpretation of the chapter on Absolute Knowing.

At this juncture, it is pertinent to revisit the contested question: what precisely constitutes Hegel's historicism? Rockmore contends that while historicism is crucial to Hegel's claims of knowledge, it is not fully developed by himself.[43] According to Rockmore, historicism stems from the contextual nature of all cognitive claims:

> If Hegel is correct, the commitment to contextualism already entails a commitment to a historical conception of knowledge, hence to historicism, which is a view about the intrinsically historical character of all first-order cognitive claims.
>
> ROCKMORE 2005, 193

Acknowledging Hegel's implicit adherence to historicism, Rockmore notes that the meaning of cognitive historicity remains somewhat "mysterious" (Rockmore 2005, 193). Consequently, he suggests that it falls upon the interpreter to "construct and evaluate arguments that support Hegel's view of the historicity of knowledge" (Rockmore 2005, 193). My objective is to undertake this task and interpret Hegel as a historical relativist. As an initial approach to this challenge, I will draw upon the reconstructions put forth by Jaeschke and Erp concerning Hegel's contentious concept of Absolute Spirit. Jaeschke

42 It is important that Fulda respects the currently emerging tendencies of analytic philosophers to interpret Hegel's absolute Knowing as a "nicht-metaphysische, 'kommunale Selbstreflexion' hinsichtlich der theoretischen und praktischen Gründe und Normen" (Fulda 2007, 386). Fulda objects, however, that these interpretations of Hegel do not discuss Spirit with regard to experience. Rather, they are dependent on the rejection of metaphysical positions that evolved within analytical philosophy. Thus, they understand Hegel "in eine[r] polemische[n] Abhängigkeit vom Zurückgewiesenen" (Fulda 2007, 386). This rejected proposition is, for example, the myth of the given. Sellars, however, introduced the myth on the basis of the earlier, analytic interpretation of Hegel and analytic philosophy's turn towards external realism. It comes down to reconstruct Hegel independently of the analytic tradition, in order to compare his model with the analytic tradition.

43 See Rockmore 2005, 194. See also Jaeschke's evaluation of this aspect (Jaeschke, Walter. "World History and the History of Absolute Spirit." *History and System: Hegel's Philosophy of History*. Edited by Robert L. Perkins, Albany: State of New York Press, 1984).

specifically examines how Absolute Spirit manifests in world history as its eternal history. Erp, moreover, argues that an inclusive notion of Absolute Knowing is grounded in the principles that individuals subjectively generate within the context of their own history. These interpretations do not signify the culmination of history but emphasize history as open to change. The insights offered by these authors will lay the groundwork for my subsequent in-depth analysis of Hegel's chapter on Absolute Knowing, wherein I will reconstruct Absolute Knowing as a concept that operates within the confines of historical relativism.

Jaeschke's observations shed light on the inherent challenges within Hegel's historicism, as he points out that Hegel did not fully elaborate on the relationship between concrete world history and Absolute Spirit. Since "the space that world history fills in Hegel's system looks comparatively small," it even occurs as an "overstatement" to interpret Hegel as a philosopher of history (see Jaeschke 1984, 102). However, as everything is a form of spirit in time, Jaeschke contemplates the "extreme position: namely that [Hegel's] system as a whole is a philosophy of history" (1984, 103). He points out that "even logic is implicitly spirit, and nature is spirit that has externalized itself," so that Hegel's whole account can be seen as historical (Jaeschke 1984, 103). The major problem with this 'extreme' interpretation, however, poses the idea of an all-embracing superspirit that works through all stages of history and entirely determines them.

According to prevalent interpretations, the super-structure of Absolute Spirit can be likened to a metaphysical mole.[44] Using the metaphor of a mole producing molehills, the notion arises that Absolute Spirit operates unseen, with all historical events and phenomena attributed to its concealed influence. However, Jaeschke raises doubts about this interpretation, asserting that it inadequately captures the true essence of Absolute Spirit. According to Jaeschke, if we construe Absolute Spirit as merely the "superstructure of several histories," we fail to grasp its full significance and depth (Jaeschke, 1984, 115). This interpretation not only contradicts the text but also risks constructing "superworlds, from which [...] a very weak glance falls back to this world and its history" (Jaeschke 1984, 115). In other words, it diminishes the significance and meaning of historical movements. More specifically, it devalues the historical agency of self-consciousness, both socially and individually, which strives for freedom and actively shapes history—an essential tenet of Hegel's philosophy. Assuming a super-structure of Absolute Spirit would render Hegel

44 Schmied-Kowarzik, Wolfdietrich. *Denken aus geschichtlicher Verantwortung: Wegbahnungen zur praktischen Philosophie.* Würzburg: Königshausen & Neumann 1999, 295.

vulnerable to dismissal by his critics, as it would provide an easy target for critique.

Nevertheless, Jaeschke writes also that the history of spirit has "without doubt [...] not only conceptual but temporal priority to the forms of history" (1984, 106). In other words, all historical events must be grounded in an original historicity, that Jaeschke calls the "eternal history of spirit" (see 1984, 106). It is true that the term of eternity can lead us easily back to the assumption of a super-entity of Absolute Spirit. Since Jaeschke, however, defends an overall historicist Hegel, it must have a more relativist meaning. In that regard, Jaeschke draws our attention to the fact that Hegel compared the eternity of Absolute Spirit to an "eternal present" (1984, 108). Following this idea, Absolute Spirit reveals itself as the underlying dialectical substance of "the process of alienation and return in itself" (Jaeschke 1984, 106). It is important for my interpretation to keep in mind that the substance of this ongoing dialectic is Absolute Spirit recognizing itself. How can this Absolute Spirit then be understood as eternal? It must be conceptualized as freely producing itself. If historical communities free themselves through self-recognition from external factors, such as contingent events in time, they can—that is the idea—fully conceptualize how they produce themselves and thereby produce themselves. This self-production can then be maintained independently of further contingent events in time. It could be called 'freedom' and defined as the goal of the movement of Spirit. There is a possibility that this relativist, self-referential framework can endure indefinitely—eternally—persisting throughout time without reaching a definitive conclusion in history.

As we have already seen in the *Lectures on the History of Philosophy:* we can find an eternal state of Absolute Spirit that is the freedom to determine itself within the limitations of its historical community. Once Spirit through the community grasps itself, and once it understands itself as a historical being that produces itself, it does not have to go beyond itself anymore. From here on—that is the major idea—it can fully conceptualize the ideal of its identity. Because of its historicity, however, it will remain a challenge to maintain this identity. On the one hand, a specific period of time ends with Spirit's self-conceptualization. In particular, the period ends that has not grasped spirit as essentially historical and producing itself. It was a time, in which Spirits contingently and disharmoniously produced extremes of unity and division. On the other hand, history continues now in another form, a form that is identical with its content. This marks the pivotal moment where the history of self-production's relentless quest for freedom sets in motion.

It is reasonable to question Hegel's project at this point. As he still attempts to mediate "metaphysics and history" (Jaeschke 1984, 115). As discussed above (part 4.1), Hegel claims a parallelism between history and logic in his *Lectures on the History of Philosophy*. It seems to imply that reason in the form of logic and reality in the form of history share the same metaphysical, ahistorical structure. We can therefore legitimately ask: does Hegel introduce only another form of an ahistorical super-structure? Since Hegel, however, did not expound his parallelism thesis, did not prove it, and even used a rather unrelated method in the concrete lectures to analyze history, Jaeschke suggests that his *Lectures on the History of Philosophy* were only about mediating historical phenomena with regard to logical thinking. With regard to the relationship between history and metaphysics, Jaeschke defines the conditions for a mediation in more detail:

> Such a meditation would not be achieved if the absolute content were present only in the separate histories of absolute spirit, and the eternal were not embedded in actual history, too, and this actual history were not elevated to the absolute.
>
> JAESCHKE 1984, 115

According to Jaeschke, the meditation of metaphysics and history depends itself on history. It implies that the logic we apply to interpret history must be bound to its occurrence in history. Jaeschke's major argument for this mediation is based on the implausibility of a super-structure and the fact that Hegel followed a different method in his concrete studies of history. Moreover, he claims that this thesis can be defended coherently with regard to Hegel's works. I will follow this suggestion.

So what does Hegel's mediation between logic and history look like? As Spirit appears in history, it is in the first place part of the various stages of world history. It has not accomplished the standpoint of science yet and has therefore no reliable logical framework to apply. The task therefore is to free the notion of spirit from intuitive, external, and contingent elements. It means to grasp its development conceptually. More specifically, we intend to understand world-history through a concept that will resemble a timeless and rational whole. It is not absurd to compare the rational whole to the logical idea—an idea that is sufficient for itself. For I follow Sedgwick, Jaeschke, Pinkard, Rockmore, and Siep, however, this logical frame is not an external, prior existing, adamantine net. Rather, the logic is itself based on history. In my view, it relies on the historically grounded standpoint of science, which is Absolute Spirit. In this sense, the *Phenomenology* can still be seen as leading us to the standpoint that then enables the thinking of essences. Naturally, these essences can and must be expressed in a science of logic—a kind of hermetic laboratory of the mind.

Here, thinking can experiment with itself as if there are no other presuppositions than Being that accomplishes identity with its thought. However, Hegel indicates in the preface of his *Science of Logic* that there are limits to his logic. Given that Plato wrote *The Republic* seven times, the *Logic* had to be rewritten at least 77. He hints at the problem that the logic is the result of a historical process that depends, at least, on a longer history of editing. Moreover, as Hegel utters concerns regarding the logic as an adamantine net in the *Philosophy of Nature*, we can suppose that the logic changes with regard to different historical demands of, for example, clarity, content or vocabulary. Afterall, logic is part of the systematic whole. Although it therefore is true that the *Phenomenology* can represent a ladder to such a logic, once we have built the ladder, we should not throw it away. Throughout Hegel's further work, for example, the phenomenological method is therefore still present in the form of how we investigate phenomena. In the concrete study of anthropology, nature, history and so forth, this method means to consider concrete phenomena and thereby to renew the cycle of reflection. Each phenomenon allows a new entry into the circle. These concrete studies express then Hegel's commitment to Spirit's eternal, historical essence of producing itself through real historical communities that experience and reflect on these experiences. The essence of Spirit is therefore not logic, but the reflective and dialectical investigation of phenomena that are experienced through a historical consciousness. In more general terms, the essence is the self-production of spirit by reflecting on all of its moments. The parts of Hegel's systematic account are for the same reason not simply gap fillers of a more essential logical framework. Their task is not only to confirm an already completed and ahistorical super-structure. With its relation to experience, the phenomenological method constantly builds new grounds for new cycles of insights. Each historical period with its own experiences therefore contributes to the open circular structure of Hegel's overall systematic and historical philosophy. From this perspective, Absolute Knowing is then only the historical moment, at which we have fully integrated all known phenomena and moments of reflection. Even though the *Phenomenology* itself occurs later only as a smaller part in the *Philosophy of Spirit,* it is therefore not merely a ladder, but as a phenomenology remains a central part of Hegel's systematic, dialectical account that takes experience seriously. In other words, we should not throw our ladder away as we might need it again when new experiences arise.

So how do we achieve Absolute Knowing through the phenomenological method? After we have recognized that natural consciousness and self-consciousness are the results of the social context, we need to begin by studying historical phenomena. Why, however, is it history that interests us? As Erp expresses it: "history, for Hegel, is the process during which spirit realises itself" (see Erp 1998, 3). It means that spirit becomes through history. Moreover, since "history is the way of how people express the principle that guides their actions" (see Erp 1998, 5),[45] we need to study people in their historical relationships to achieve an adequate concept of Spirit. This is because as people build societies on the ground of common principles, they form communal spirits. As their societies change, spirits change, which in turn creates a succession of different historical societies, so-called world-spirits. Spirit is then the substance of these societies, while the substance is the subjective action of people within historically grown communities. The goal of Absolute Spirit is to understand how it is the result of these actions and, at the same time, guides these actions. In this inter-dependent structure, Spirit can find the possibility to recognize itself: a structure of grounding itself through itself within its history, i.e., historicity.

Once Spirit understands itself, it understands how the contingent succession of world-spirits can be conceptualized as one history of producing itself. Subject and object would then coincide and thus fulfill the formal and standard truth-criterion of a correspondence theory. If we can achieve this goal of self-understanding, an objective standpoint in history would be reached that, in accordance with Rockmore's interpretive demands, remains within the boundaries of an objective, yet relativist historicism.

So far, we have defined the formal goal of Hegel's *Phenomenology*, i.e., the integration of all experiences and reflections in one identity. In order to understand how Absolute Spirit works concretely, however, we also have to understand first the principles of people. These principles can be very different. Each person organizes their life according to a principle or principles. For example, we can decide that the principle of our life is to make people laugh, while another person may decide that their principle of life is to gain wealth. As much as we can decide to be the witty, the desiring, the honorable, or the wise person, as much such principles can be the guiding principles of societies. We shall be honorable, a union of soldiers, or a union for graciousness, and so forth. These principles are not yet absolute, meaning that they are not yet reflectively acknowledged. Rather, they come by intuitive insights and their reason for adoption is not always clear. Such a principle can be, for example, a stoic or a skeptical attitude. Because of their not further reflected origin, however, they are limited. This means that at some point in time, they will contradict themselves. For example, a libertarian claims that property rights and freedom of speech are undeniable principles. Freedom and property rights,

45 Erp, H.V. "The End of History and Hegel's Conception of Modernity." *Ideas Y Valores* 47 (107), 1998. http://www.revistas.unal.edu.co/index.php/idval/article/view/21868.

however, can conflict. If, for example, students want to express their opinions at a private university, the question emerges of whether the principle of property rights or the principle of free speech is more important. If the practical limitations of a concept become too pressing, people will change their principles. The libertarian might acknowledge at some point that his principles are incomplete. As societies in the form of nation states undergo the same procedure, it can ultimately change the course of their world history and bring forth a new historical shape. The principles of people become principles of societies and propel a historical dialectic. However, as long as the principles are based on intuitive insights, include external elements, are theoretically abstract, and formulate extremes, this historical dialectic will remain a blind struggle within history. As Erp argues, the "spirit of a particular people will always be finite only" (Erp 1998, 5). In other words, their self-understanding will be limited and lead to contradictions. We therefore need to develop a version of Spirit that reaches its maturity in self-knowledge.

Given this general idea of how principles are produced by real, historical individuals, we can now attempt to understand the "development of the idea of the spirit as such" (Erp 1998, 6). For the *Phenomenology,* I understand this as the movement of natural consciousness towards science which is the Absolute Knowing of Spirit. Since history is the result of the actions of people, the former, insufficient principles occurring in world-history are then moments that Absolute Spirit has to reflect on. People can be aware of their Spirit, but first only in a convoluted form. Since they cannot yet fully conceptualize their own principles and thus cannot understand their actions, their long history produces various extremes. Absolute Spirit can therefore be seen as a return from these extremes. It avoids extremes by eliminating the non-rational elements. It conceptualizes the finite shapes of spirit and understands for why they lead to extreme positions that conflict. The central idea is that in reflecting on itself and its inner historical dialectics, it can establish an identity of differences—differences that otherwise would fall apart.

The central aspect of the phenomenological method is to overcome intuition and thereby to initiate true conceptual thinking. If we eliminate all intuitions, we will be left with the substance of all the former historical development. Absolute Spirit as a result of this elimination will still be rooted in history but makes transparent its own emergence as a history-producing force. As its goal is the complete conceptualization of this emergence, it will, indeed, appear to us as "a rational totality" (Erp 1998, 6) that can maintain itself. Expressed in the religious representation, for example, it would demonstrate itself as "a matter of divine providence" (Erp 1998, 6). Without religious intuition, however, the rational totality is conceptualized as the result of free, historical people who give themselves principles within their historical limitations. The rational totality therefore only *seems* to be a hidden power that comes in the form of "a philosophical apriori" (Erp 1998, 6). According to the interpreters that I follow here, however, Hegel is neither a philosopher who defends apriori principles nor does he defend an instantiation of God in Absolute Spirit. Rather, he attempts to acknowledge the freedom of people to determine their own history. I therefore interpret the "rational whole" of spirit as the gradual realization of the people's freedom within the boundaries of their historical communities (see Erp 1998, 6). As this project of freedom is based on historical, conceptual reflection, spirit determines its own substance a posteriori (see Erp 1998, 6). Its criterion of success is the full conceptualization of how it produces itself.

Let us come back to the controversial term of the eternal here. What is it that makes Absolute Spirit as the completely conceptualized freedom of the historical self-production eternal? It is eternal because it fully conceptualizes its essence. It frees itself from intuition. In this form, historical Spirit becomes what it is: a history-producing and thereby self-producing entity. Although the historical boundaries cannot be superseded, it accomplishes the eternal truth of freely making its own history within its historical limits indefinitely. If it is successful in defending its identity, it can accompany all further, historical change within these historical limits.

The next controversial term in that regard is the conception of Absolute Spirit as 'the end of time'. Once Absolute Spirit knows itself, a specific time of development comes to an end. Erp describes this as the rational end of Spirit's development in relation to world-history:

> A philosophical concept of world-history can only be developed from the moment that world-history itself has reached the spiritual result that can be conceived of as its rational end. Therefore, world-history is itself a special or concrete totality of time; it is time that has come to its end.
>
> ERP 1998, 6

It means that Hegel's concept of Absolute Spirit is the rational end of temporal development. All former development has led Absolute Spirit to knowing itself as a self-producing concept. World-spirit was guided by convoluted principles and has reached its limits within a specific time. Absolute Spirit emerges then as a new form and content. It conceptualizes all former events as a history of its self-production. With Spirit's transition into the concept, time loses its significance for its further self-understanding. Time does not disappear, but we could say that with regard to the fully conceptualized identity of Absolute

Spirit time loses its relevance. The end of time therefore does not imply the end of history. From here on, we know only how to locate our position within history as subjects of a self-producing history. We have made ourselves transparent to ourselves with regard to our history. This truth of a self-producing identity within history is Absolute Knowing.

My major point concerning Hegel's distinction between time and history is therefore: Absolute Knowing reflects on all former moments of its historical becoming. It determines itself as a historical, rational whole of this becoming. At the same time, it rejects time as a ceaseless infinity and repetition of the same. It now opens itself for the free development of its concept. Since Absolute Knowing is, however, also the result of its becoming, it remains within the framework of history. We leave time only as a succession of contingent events behind, and thereby, if possible, "free the notion of history from the notion of natural time" (Erp 1998, 7). We are not suddenly outside of time, but we now conceptualize ourselves rather in the concepts of history than in the intuition of time.

With regard to the end of time, Erp explains that the final stage of Absolute Knowing closes "the history of the world, but without the necessity of being understood as its historical end, that is to say, as an end-point in time" (Erp 1998, 12). The end of world history is an achieved actuality of self-knowledge. In other words, we can interpret Absolute Knowing as concluding a historical movement of Spirit and preparing us for another richer, conceptual, and historical movement of itself.

Other philosophers believe that "the dialectical closure of philosophy excludes a real future development of spirit" (Erp 1998, 14). Yet, according to Erp, it is flawed to compare "Hegel's concept of spirit with some complex and abstract representation of a subsistent entity, separated from our conception of ourselves" (Erp 1998, 14). I have already noted that such an interpretation of Hegel's Absolute Knowing has influenced the analytic rejection of Hegel. Some Hegelians, indeed, defend a view of a religious super-substance that determines all our actions. They believe that Absolute Spirit expresses the identity of reason and reality so that we can understand the whole course of history including its future. In other words, history's destiny is God's plan. Russell's philosophy was reasonably directed against these, we should say, philosophical religions. In a sense, also Habermas criticizes these religious Hegelians. His critique of Brandom is justified because he finds in his conceptual realism a similar model that replaces God with a super-concept of semantic reality. Once humans realize their relationship with a mind-external, semantic reality, its conceptual structure will, according to Brandom's model, dictate their path into their future. Brandom was guided here by the idea that Hegel, in order to avoid the problem of an epistemological hiatus, defends an identity between mind-external reality and subjective cognition. The assumption of such an identity is a controversial topic of Hegelian philosophy. Erp addresses the major criticism of this identity:

> Hegel's notorious statement about the identity of actuality and reason sounds unbelievable to most of our contemporaries and even as blasphemy for some. Fifty years after the Second World War, it has almost become a commonplace to confront the pretensions of the Hegelian philosophy of history with the 'truth of Auschwitz'. Auschwitz would be the absolute refutation of Hegelian thinking of totality, with its identification of history, reason and providence.
>
> ERP 1998, 15

As Erp points out, it is, indeed, absurd to claim that the course of our history was rational with regard to Auschwitz. The identity of reality [das Wirkliche] and reason is, therefore, a controversial topic. But does Hegel say that cruelties like Auschwitz are a necessity for reason to realize itself? It is true that Hegel's identity of reality [das Wirkliche] and reason is implied in Absolute Knowing. The question, however, is: what does he mean by this identity? I argue that it means something different from what critics of Hegel usually assume.

Hegel introduces the thought of the identity of reality [das Wirkliche] and reason also in the *Science of Logic*. He maintains that reality must be rational since otherwise it could not be known. In my view, it is crucial that he does not mean reality as a mind-independent reality, but a reality that we empirically experience and that involves our perspective—a reality that is for us. His argument in the 'Objective Logic' is that reason, as the subject's capacity of reflection, recognizes that it needs a notion of reality. Otherwise, it would fall back into mere subjectivism according to which the subject is limited to only internal concepts. It would never achieve objectivity. It needs a concept of reality. Nevertheless, this notion of a necessary reality depends on the subject recognizing this necessity. In a typical Hegelian twist, the subject therefore produces an in-itself that is for-itself. Here, I argue further that since the subject is not a stable entity, but becomes through forms of communal spirit, reality that is for us must evolve with the dialectical movement of the subject. If we then say that the rational is real and the real is rational, it only means that both undergo the same changes within the tides of history. The problem is therefore not the identity of reason and reality. Rather, it lies in the question of whether there is a last and final stage of the development of subjects within history. It would bring us to a further neuralgic point in Hegel's philosophy, the question of whether Hegel has a closed and final system.

There are some indications that Hegel did not see his work as a completed system. In a letter to Victor Cousin about his second edition of the Encyclopedia in 1827, he writes, for example: "This book is only a collection of claims" (Hegel 1954, 169).[46] "A collection of claims" lacks the completeness that is required for interpreting it as a fully developed system. And also Schalhorn emphasizes that Hegel understood his *Encyclopedia* merely as a guideline, which does not give account to a complete system (Schalhorn 2000, 14).[47] Rather, this guideline has to be *taught* in lectures. According to Schalhorn, Hegel expressed this restriction in all three editions of 1817, 1827, and 1830 (see Schalhorn 2000, 14). In other words, Hegel rather relates his systematic approach to the idea of teaching, guided by a desire for truth and the nature of human beings to learn (to actualize their potential as Hegel expresses it in the *Lectures on the History of Philosophy* in Aristotelian words).

Like Jaeschke and Erp, I therefore argue that Hegel's idea of Absolute Knowing must be linked to his historicism. My idea is that Absolute Knowing occurs only in history and is thus limited to its historical emergence. In that regard, my discussion focuses on Habermas' interpretation concerning the supposed annulment of time. It concerns a reinterpretation of this passage in the *Phenomenology*:

> In the Notion that knows itself as Notion, the *moments* thus appear earlier than the *filled* [or *fulfilled*] whole whose coming-to-be is the movement of those moments. In *consciousness*, on the other hand, the whole, though uncomprehended, is prior to the moments. Time is the Notion itself that *is there* and which presents itself to consciousness as empty intuition; for this reason, Spirit necessarily appears in Time, and it appears in Time just so long as it has not *grasped* its pure Notion, i.e. has not annulled Time.
>
> HEGEL PoS §801, 487

So I ask again: what does it mean to annul time? The moments necessary for achieving Absolute Knowing precede the result of Absolute Knowing in a succession of temporal events. However, as these moments of Absolute Knowing move in time, they are still external to Spirit. The goal is therefore to

46 "Ce livre n'est qu'une suite de thèses"(Hegel, G.W.F. *Philosophische Bibliothek 237*. Hamburg: Meiner, 1954, 169, translation by me).

47 Schalhorn, Christof. *Hegels enzyklopädischer Begriff von Selbstbewußtsein*. Hamburg: Meiner, 2000, 14.

conceptualize Spirit's movement in time without presenting it as a contingent, temporal succession of stages.

Hegel presented the basic idea for this conceptualization of time in the preface of the *Phenomenology* with regard to the example of a plant. In time, the fruit does not appear before the bud (see Hegel PoS §2, 3). The whole plant is a succession of stages. Yet, neither the parts of this whole nor the whole of the plant are the plant. According to Hegel, it is the philosophical challenge to present the plant as the ultimate result of all of its moments and the whole of its being. In other words, the becoming and the result both belong to a single moving essence. With regard to Absolute Knowing, it is similar. If we presented the result of Absolute Knowing as a dead, teleologically preconceived goal, then we would posit it as a lifeless result. Consequently, the core of Hegel's analysis cannot be a universal result. Moreover, it is also not the individual moment of one historical instantiation. Rather, it is about freeing both the individual and the universal from their contradiction in time by demonstrating that both are necessary moments of one developing Absolute Spirit. We must grasp the process itself that produces the intuition of the individual moment of time and the universal concept. Since this process can neither be after or before, nor simply now, time must be annulled. Absolute Knowing itself is not in time, but it must be grasped as the conceptual ground for the conception of time. In this way, Absolute Knowing, however, remains relative to its historical moment, in which it occurs and according to which it is conceptualized.

Hegel's discussion of Absolute Knowing is difficult to understand since it denies the individual, abstract notion of time. It denies temporality, that is time as vanishing in the fractions of seconds, and it denies the universal notion of time, time as a flat infinity that has no elements. It therefore only seems as if Absolute Knowing denied or rejected time, while we actually attempt to sublate it in the higher concept of history. Hegel, therefore, writes on this relativizing of time:

> Time, therefore, appears as the destiny and necessity of Spirit that is not yet complete within itself, the necessity to enrich the share which self-consciousness has in consciousness, to set in motion the *immediacy of the in-itself*, which is the form in which substance is present in consciousness; or conversely, to realize and reveal what is at first only *inward*.
>
> PoS §801, 487

Westphal remarks that "the temptation is all but irresistible" that Hegel refers to a "transtemporal, super-human Absolute" as a "timeless knowing" (see Westphal 1979, 219). Yet, as Westphal further observes, Hegel is not concerned with time's "duration or succession" (Westphal 1979, 220). He rejects the problematic externality of temporal relations. Only in the religious consciousness, which is still a consciousness of representation "the unity of human and divine is conceived of as a temporal event" (Westphal 1979, 220). Instead, Hegel suggests a non-representational solution that abandons the subjective form of thinking that is bound by categories or intuitions such as time. Thinking really achieves its substance as the standpoint of the historical subject. This Absolute Knowing occurs when the knower knows its knowing as a particular result bound up in his or her particular, historical moment. From this, it further follows that Absolute Knowing is relative to its self-knowledge, in particular to its historical moment. This historical moment can only be with regard to its earlier historical becoming, the succession of stages that led to it. However, before I can come to the concrete analysis of this succession—as Hegel develops it in the last chapter of the *Phenomenology*—I need to discuss Hegel's formal concept of the Absolute.

3.2 *The Formal Absolute*

In the following passages, I briefly discuss Hegel's concept of the absolute. 'Absolute' usually stands in for a concept that is independent of anything else, is final and can serve as a principle (see Rockmore 1997, 179). In the Kantian philosophy, the absolute was posited as a transcendental condition for knowledge, only having indirect access to the absolute. In other words, the absolute has to be posited outside of ourselves as a noumenon in order to achieve an understanding of our otherwise incoherent concepts (see also Rockmore 1997, 179).

Hegel's perspective emerges from a prolonged discourse within the German Idealist tradition, wherein the noumenon is eventually discarded as a transcendental precondition as it is based on a questionable extrapolation. Hegel's objective is to introduce the notion of the absolute in a different manner. The central challenge at hand revolves around the seemingly insurmountable predicament of reconciling the ontological distinction between subjective agency and the external object situated outside the subject.

Unlike Brandom, the tradition that Hegel arises from pursues an approach that is internal to the mind. To satisfy the standard truth-condition, which requires the subjective, conceptual activity and the determinate object to align, we must locate a singular object that can be contemplated. According to Hegel's standpoint, the object sought after is Spirit. Spirit possesses the ability to embody subjective, conceptual activity while simultaneously manifesting itself as the object of said activity. In this manner, the activity and the resultant object can be identical, thereby fulfilling the fundamental condition of a correspondence theory of truth. If we can demonstrate that Spirit, as a subjective activity, can comprehensively understand itself as the acting object, it would be independent of external dependencies. This understanding could then be termed absolute knowledge, resulting in the manifestation of Absolute Spirit.

The sought unity of Spirit, however, cannot be an empty and abstract formula like 'I'='I', an idea that Fichte committed to. Instead, the sought unity must also be recognized in its process. In other words, the standpoint of Absolute Knowing has to be achieved as a unity of being and becoming. Hegel commits to this view of an identity already in his *Differenzschrift* and also defines the form of the investigation that is related to his solution:

> But if the Absolute, like Reason which is its appearance, is eternally one and the same—as indeed it is—then every Reason that is directed toward itself and comes to recognize itself, produces a true philosophy and solves for itself the problem which, like its solution, is at all times the same. In philosophy, Reason comes to know itself and deals only with itself so that its whole work and activity are grounded in itself, and with respect to the inner essence of philosophy there are neither predecessors nor successors.
>
> HEGEL 1977b, 87[48]

The primary objective of philosophy, according to Hegel's philosophy, is to attain the Absolute, where "subject and object, knower and known" become harmoniously fused (Rockmore 1997, 182). This unification is pursued through the process of reflection, which necessitates that reflection arises from its underlying substance. Crucially, the substance must be demonstrated as the very essence of subjective reflection, substantiating its connection to the content at hand. In other words, reflection must stem from its substance, and substance must be demonstrated as the content of subjective reflection.

If the unity of subject and object constitutes the purpose of philosophy, it signifies more of a necessity rather than a predetermined outcome. Moreover, it suggests that in order to achieve unification, there must initially exist a distinction that requires resolution, as philosophy emerges from such differentiation. The process of reflecting upon the Absolute unveils the imperative

48 Hegel, G.W.F. *The Difference Between Fichte's and Schelling's System of Philosophy: An English Translation of G. W. F. Hegel's Differenz des Fichte'schen und Schelling'schen Systems der Philosophie.* Edited and Translated by Walter Cerf and H.S. Harris. Albany: State of New York Press, 1977b.

of transcending the subsequent contradictory differences in order to apprehend them within a comprehensive framework of unity. In his epistemological exploration of Hegel's *Phenomenology*, Rockmore explains that the notion of the Absolute holds the potential to offer such a unifying force:

> The idea is, then, twofold since it expresses the identity of the concept and objectivity and further expresses the relation of the subjective and objective poles, or the concept and objectivity as immediately given. This definition is now said to be absolute and, as such to supersede all previous definitions of the absolute.
>
> ROCKMORE 1997, 184

In other words, the task of the Absolute is to provide unity between an experiencing subject and the object of its experiences. For Rockmore, Hegel holds on to the Absolute "as an ultimate principle, which is independent of all further principles" while this constitutes "monism" (Rockmore 1997, 184). It is a monism because the Absolute unifies all former differences.

I would like to emphasize that Hegel's 'monist' approach significantly diverges from Brandom's semantic account. Hegel's monism serves as an epistemological imperative for attaining knowledge but does not encompass assertions about a monist ontology. Rather, the primary objective is to comprehend objectivity without presupposing the existence of an external reality. It is an epistemological undertaking.

In the present historical context, the only object capable of embodying the sought-after unity, resembling the desired Absolute, is, in my view, reflective reason in the objective manifestation of Spirit. By incorporating Spirit, Hegel ultimately moves beyond the pursuit of abstract conditions of knowledge (Rockmore 1997, 184). Instead, he examines how knowledge emerges as a phenomenon wherein a subject relates to itself through the presence of others within historical communities.

However, the Absolute remains an ideal that provides an epistemological framework for unity while simultaneously serving as the underlying reason for the existence of these reflective communities. Hence, the Absolute, as the object sought by communities, and the subjective activity of community-building must be understood as one and the same: Spirit. In other words, Spirit is not contingent upon a mind-external reality. From my perspective, Absolute Spirit represents the most viable solution to the epistemological quandary of knowledge generation that we currently grapple with. It indeed entails an ontology encompassing theories about the experienced world. However, these theories are subject to change with evolving historical conditions, including the potential for change within Spirit itself.

In the subsequent part, I will explore Hegel's resolution of Absolute Knowing in detail. As mentioned earlier, Habermas critiques Hegel's reintroduction of an eternal Absolute, dismissing it as dogmatic metaphysics. Nevertheless, Hegel's concluding paragraphs on history and the transcendence of time within Absolute Knowing reveal that historicism is a vital part of Hegel's framework.

4 The Achievement of Absolute Knowing

We shift our focus to the specific text of Hegel concerning Absolute Knowing, particularly the final chapter of the *Phenomenology*. This chapter is arguably one of the most challenging in the history of philosophy. Hegel endeavors to explain the entire history of consciousness reflecting on itself and grasping itself in various forms of spirit, ultimately condensing this into the concept of the concept. Reflecting on my own writing process, I believe I have finally provided a clear outline that accurately reflects Hegel's thought. However, I must admit that editing it numerous times over the years, each revision cycle has uncovered new ambiguities. Every time I conclude that my interpretation is clear, a new round of editing reveals further uncertainties. I therefore hope readers recognize the inherent challenges of this chapter and view my interpretation as a contribution to the long history of efforts to clarify Hegel's thoughts, rather than judging it by an unattainable standard of absolute clarity. While this may reinforce the stereotype that Continental philosophy is obscure, I hope it is met with the awareness that we are engaging in the underlying problem: how to understand the object in front of us objectively, despite our understanding being influenced by the historical context that we, as self-conscious beings in a historical community, created and that creates ourselves. It is the difficulty of clarifying the absolute context once and for all that has guided our understanding from the beginning and continues to shape the very attempt of clarification.

Considering that this chapter is arguably one of the most challenging texts in the history of philosophy, I will begin by summarizing the key points within each paragraph. Subsequently, I will analyze each paragraph, extracting the central elements that culminate in the historicity of Absolute Knowing. In §788 and §789, Hegel briefly addresses the issue of spiritual identity within religion. He explains that while religious consciousness has an intuitive sense of self-consciousness's unity, it struggles to conceptually grasp the implications of this identity. From §790 to §795, Hegel summarizes the preceding development of Spirit within religion. He elucidates that the movement of consciousness propels it beyond its immediate state, giving rise to various forms of self-consciousness. Ultimately, all diverse forms of worldly knowledge converge in religious knowledge.

Religious knowledge initially presents itself as a contradiction between the beautiful soul, the good, and its realization, the evil. The reconciliation of this contradiction gives rise to an implicit concept of forgiving. Yet, the fundamental challenge remains: substance has not yet become subjective, and the subject has not yet become substance. Even in its highest form, religious communities can offer only intuited solutions, unable to fully rationalize and thus resolve the problem of unity (§796). According to Hegel, philosophy surpasses this limitation by comprehending spirit through the Concept (§797).

In §798, Hegel elaborates on the final and ultimate shape of Absolute Knowing, referred to as 'the Concept'. However, the unity within this concept remains purely formal. Recognizing this, §799 explores the role of individuals who experience Absolute Knowing as a personal act. Nevertheless, human beings primarily grasp the different stages of spirit and its development only through intuition. In §800, Hegel demonstrates that the task of science is to systematically navigate these subjective stages of knowing to attain Absolute Knowing. However, a further fundamental challenge cannot be simply resolved: the substance exists prior to subjective comprehension, indicating its external nature to thought. §801 highlights the incomplete nature of spirit's manifestation, still containing elements of contingency. Hegel proposes that time, as the residual aspect of intuition, needs to be annulled. It is important to note once more that Habermas interprets this annulment of time as a denial of historicity. However, as I will argue, it is a necessary step towards attaining historicity.

In §802, Hegel delves into the transformation of "Substance into Subject" (Hegel PoS §802, 487). This transformation elucidates the concrete relationship between cognition and experience. Experience serves as the starting point, involving subjects and establishing an intuitive connection to substance. §803 introduces the conceptual argument concerning Absolute Spirit as an intuited substance within history. Specifically, Absolute Knowing emerges as the culmination of Spirit's endeavor to grasp itself through various historical forms. Through a historical analysis, Spirit manifests as a result of human actions across different historical stages, evolving and assuming complex identities that are only intuited.

Hegel summarizes the prevailing historical moment by highlighting that previous philosophical positions assumed the existence of an Absolute beyond thought. To resolve this predicament, Hegel draws upon Spirit's capacity for reflection, enabling self-differentiation (§804). He develops a final version of Absolute Knowing, grounded in the concepts of negation and reconciliation. These concepts transcend time and embody the unity of all historical processes within their inherent negativity. In §805, Hegel describes this solution as the existence of the Concept, where the Concept no longer disintegrates into separate notions of truth and knowing. In this context, Hegel foreshadows his idea of the *Logic* and presents a unity independent of time. However, it is crucial to note that this envisioned science of logic does not come to a standstill or represent a stable, ahistorical endpoint, as Habermas' interpretation may suggest. Rather, this logic reintroduces us to the cyclical nature of Hegel's systematic account by returning to nature through externalization. Accordingly, §806 introduces the notion that knowing the limit of an object signifies being already beyond that limit. In the case of Spirit, this knowing entails the 'sacrifice' of Spirit itself. Absolute Knowing, in its highest logical form, freely externalizes itself. Firstly, it externalizes itself in Nature as its immediate manifestation. Secondly, it externalizes itself through Spirit as History. While §807 explores nature as an externalization within the realm of contingent time, §808 further develops the concept of history, ultimately eliminating the last vestiges of contingency. However, science and history are two interconnected aspects that must be understood together in our present historical moment as culture [*Bildung*]. Human beings possess knowledge of culture, and since knowing inherently implies surpassing, it will always engender new beginnings within the realm of culture [*Bildung*].

The concept of 'history' is not mentioned before the last four paragraphs of the *Phenomenology*. This, however, should not lead us to overlook the significance of history within the entire *Phenomenology*. The focus on history is evident throughout the *Phenomenology* with its focus on a developing history of spirit, and the concluding paragraphs serve to reinforce its importance. Moreover, the final two paragraphs explicitly illustrate that history does not culminate in the idea of Absolute Knowing; rather, history is intrinsically linked to the process of externalizing spirit through culture [*Bildung*]. Absolute Knowing involves comprehending one's own cultural journey, which, in turn, transforms the very fabric of that cultural process. Thus, Absolute Knowing is an ongoing and perpetual endeavor within the realm of history, requiring continual achievement with each new cultural beginning. The history of Absolute Spirit thereby remains an active and evolving process, far from reaching a final destination—an end of history.

In the following points, I will provide a more detailed exploration of these steps, elucidating their significance and drawing connections to the problem of history that Hegel addresses in the final four paragraphs of the chapter on Absolute Knowing.

4.1 *The Exposition of the Problem of Spirit*

Following Kant's premise that reason can only apprehend what it produces (see Rockmore 1997, 181), Hegel focuses on reason's reflective capacity. However, the process of self-knowing, while compelling in its potential, is not a straightforward unity, as, for example, in the philosophies of Descartes or Kant. Consciousness cannot immediately bridge the gap between truth and knowing, since it can only perceive itself as an object, unable to grasp its true essence as a subject. Hegel, therefore, suggests that true self-understanding comes through immersion in a historical human community, recognizing how the community's objects stem from their own activity and, in doing so, how they create selves (see PoS §788, 479). In essence, Hegel replaces the theoretical notion of self-consciousness with the practical concept of a historical and social Spirit. The new challenge, therefore, is to demonstrate that consciousness recognizes itself within the stages of its spiritual and historical becoming within a community. By acknowledging itself as a co-creator of these communities of rational individuals, it establishes an identity between subjective activity and its produced object, i.e., Spirit. It comprehends itself in the action of shaping itself through a community. Moreover, the individuals of this self-made world can truly call themselves rational (see Harris 1997, 709),[49] as they elucidate how their rationality grounds themselves in the context of their own historical community. As this self-making of Spirit is free from externality, their standards of how to achieve themselves can serve as a truly rational foundation for their community and for science.

At this juncture, Hegel faces a significant challenge in achieving this envisioned community, which aligns with Brandom's critical perspective on representationalism. In the final stage of religion, there is an inherent limitation in bridging the gap between subject and object, as it fails to provide a comprehensive clarification of its community standards. Religion relies heavily on intuition and has not yet attained complete rationality.

However, it is essential to note that Hegel does not endorse a position akin to Brandom's conceptual realism. Brandom's reliance on mind-independent entities as the foundation for communication would reintroduce an external and unexplained element into rational discourse. Conceptual realism aims

49 Harris, H. S. 1997. *Hegel's Ladder. Volume II: The Odyssey of Spirit.* Indianapolis: Hackett Publishing.

to establish a connection to an object that has not yet captured all of its elements as subjective creations. This approach keeps substance distinct from the subject, introducing an unexplained aspect that falls beyond its theoretical framework. Consequently, it perpetuates an irrational element, namely, the existence of mind-external objects.

Instead of commencing with the object, we must remain on the side of the subject. Yet, Hegel's objective of subjectivizing substance presents a further challenge: it appears to diminish the objective value, reducing it to a state of nothingness, leaving only the subject without a world. Hegel, however, argues that we must perceive this nothingness as a positive outcome. As he puts it, "self-consciousness knows the nothingness of the object [...] because it externalizes its own self," ultimately gaining itself as a positive result of its negating (PoS §788, 479). In essence, self-consciousness demonstrates its capacity to create and shape something out of nothing. It reveals how the subject forms objects through externalization, enabled by "the indivisible unity of being-for-self, whereas self-consciousness is also, at the same time, in-itself" (PoS §788, 479). However, even acknowledging this, consciousness cannot simply and fully access its creation process, and attain complete self-consciousness. The understanding of this process remains external and thus distorts it. True self-production cannot be reconstructed in theoretical terms but needs the involved historical subject. In other words, consciousness must actively undergo the experience of alienating objectification and reconciliation with itself in a society. At this point, Hegel asserts that consciousness can achieve this "communion with itself in its otherness as such" (PoS §788, 479). It means that self-consciousness can actualize itself through societal processes where it meets itself as the other and achieves reconciliation with this other.

Hegel's concept revolves around our self-discovery through self-externalization in society, which reaches after multiple stages a preliminary culmination in revealed religion and its concept of forgiveness. However, in revealed religion, the synthesis of substance and subjectivity is reached solely through intuition without complete conceptualization. In other words, revealed religion merely "affirms" consciousness' "spiritual essence" (PoS §789, 480), but it does not yet understand it. So how can we subjectively grasp the unity of self-generating consciousness and its object of consciousness within the context of human history? Hegel's perspective suggests initiating with natural consciousness and its created object and gradually advancing through the stages of Spirit toward his preliminary formal solution: The Concept. In its practical application then, the Concept can transcend historical boundaries, surpass mere intuitive identities, and ultimately achieve a rational understanding of its own historicity—its own becoming and being.

In order to make this movement of the concept transparent, we circle back to the starting point—natural consciousness—and explain the problems emerging from it that The Concept is supposed to solve. For natural consciousness, the produced object is not a mere fantasy; it must possess objectivity. However, there should be no presupposition of anything external to us, such as a specific metaphysical reality (as Brandom suggests). Once more, Hegel recaps the stages of consciousness experiencing objects: first, "the object is in part immediate being or, in general, a Thing-corresponding to immediate consciousness; in part, an othering of itself, its relationship or being for-an-other, and being for-itself, i.e. determinateness-corresponding to perception; and in part essence, or in the form of a universal-corresponding to the Understanding" (PoS §789, 480). This means that, firstly, in the case of sense-certainty, the object is sensed as existing external to consciousness, appearing as an entity bearing an immediate and seemingly complete truth. However, consciousness cannot linger in this immediacy because it cannot ascribe any meaningful distinctions to the many 'this's' and 'nows' that a single universal 'this' or 'now' encompasses. This realization unveils a dialectical interplay between the object and its perception. The second stage of cognition emerges: consciousness reveals the thing to be a "determinateness corresponding to perception" (PoS §789, 480). This alteration in the relationship between unity and multiplicity results in the object's paradoxical nature: it is simultaneously one entity and composed of many individuals. Consequently, as the third point, the dialectic of the object oscillating between unity and multiplicity must be guided by an "essence" comprehensible "in the form of a universal corresponding to Understanding" (PoS §789, 480). Consciousness encapsulates the totality of these three moments. None of these moments can exist independently of consciousness. Thus, in an initial act of self-awareness, consciousness reveals that the object mirrors its own nature.[50] Consciousness is the essence of the object.

Hegel now explores this essence of consciousness, explaining its conceptual activity as a totality, akin to a syllogism. This process involves the movement from the universal to individuality and the reverse movement from individuality to the universal (see PoS §789, 480). However, this self-consciousness remains also only formal, framing the problem of self-knowledge within the formal context of the relationship between the universal and the individual,

50 Harris reconstructs these passages strictly with regard to the appearances of God in the revealed religion (see Harris 1997, 714–715). Technically, this is accurate with regard to the text. I hold, however, the position that, overall, Hegel wants to present his epistemological position. For this reason, I attempt to avoid too many references to God and its appearance in religion.

a concept reminiscent of Hegel's *Logic*. Logic, in essence, aims to bridge the gap between the individual and the universal through syllogisms. It is, however, important to note that 'universal' and 'individual' are abstract terms that do not fully encapsulate their essence either. Within the context of the *Phenomenology*, Hegel rather demonstrates our capacity to establish a concept that is not reliant on external terms. Only by shedding external terms like 'individual' and 'universal' can we attain objective knowledge concerning our history, ultimately enabling us to understand ourselves within the historical moment through 'the Concept'.

My point is that self-knowledge, derived from elimination, is not based on an unrelated or detached thought but rather stems from a phenomenological inquiry into our experiences within culture and history. This process can be formally presented as a transition between the universal and the individual, utilizing logical terms. In this logical context, the process is perceived as resembling the internal subjective conceptual activity of 'the Concept'. However, we must also first comprehend this conceptual activity within a specific historical moment. Before delving into the purely logical movement of our conceptual activity, we therefore need to demonstrate how consciousness can grasp itself as self-consciousness through its concrete manifestation in history. In this regard, Hegel notes that so far "this Knowing is to be indicated only in its process of coming to-be, or in the moments of that aspect of it which belongs to consciousness as such, the moments of the Notion proper or of pure knowing in the form of shapes of consciousness" (PoS §789, 480). Fundamentally, our goal is to provide a comprehensive explanation of the shapes of consciousness that occur in society and to ultimately provide a unifying notion for them that does not rely on external terms. This happens by examining history, while we steadily draw by virtue of negation nearer to Absolute Knowing in the form of 'the Concept'. Our historical exploration will then yield the conceptual form of 'the Concept', capable of reconciling the contradictory elements emerging within collective consciousness, i.e., Spirit. The genesis of the concept of the concept arises from our rejection of any external, preconceived elements from our historical context, and offers a non-intuitive, fully rational solution. Nonetheless, as I will illustrate, it remains interwoven with both culture and history. With this objective in mind, we embark on a renewed investigation to reconstruct the inherent contradiction that the concept must bridge—an internal challenge arising from the need to unify consciousness and self-consciousness.

4.2 *The Unification of Consciousness with Self-Consciousness*

Let us recap the problem of self-consciousness: as consciousness aims to comprehend its experiences, it evolves into self-consciousness, the essence underlying these experiences. However, at this stage, consciousness primarily apprehends itself through external facets, constraining its perception to view itself as an object rather than as an active subject. In response to this challenge, Hegel contends that self-consciousness must identify itself with others in the mode of mutual recognition. Mutual recognition evolves in ethical communities where collective consciousness, referred to as 'Spirit', gradually reconciles the divergences that arise from society's imbalances. It is important to emphasize that these unities of Spirit remain in a state of perpetual evolution and bear the indelible mark of history. This entire process finds its origin in the task of unifying consciousness and self-consciousness, a challenge that Hegel now scrutinizes in deeper detail.

There are again three stages in the evolution of consciousness becoming self-consciousness: the immediate, the perceived, and the understanding. However, each of these stages falls short due to their reliance on external factors. In the initial stage, where consciousness seeks immediate recognition, we form judgments that treat the 'I' as an object. Here, self-consciousness is perceived solely in its external aspect, regarding the 'I' as a tangible entity. For instance, this perspective was evident in phrenology, where the skull's structure was believed to offer insights into the 'I'. However, this viewpoint, grounded in the idea of the immediate presence of consciousness in the world, represents an extreme 'materialist' position.

In the second stage of perception, we swing due to contradictions to the opposite extreme, resulting in the notion that "the thing is I" (PoS §790, 481). In this stage, our attention is directed exclusively on the subjective facet of consciousness. Here, the immediate object is perceived in relation to consciousness' utility, continually shifting depending on the observers and their intended purpose for the object. All objects are filtered through the lens of consciousness, and their understanding is thus contingent on the reasons cultivated within communities of consciousness.

The recognition of consciousness' essence oscillates between the immediate perception of the 'I' as a lifeless object and the perceived utility of these appearances for a community of 'I's, lacking stability. This oscillation leads to a third stage of understanding, where moral self-consciousness emerges. In moral self-consciousness, the 'I' is neither a mere objective fact nor a mediating subject of random purposes. Instead, consciousness realizes that everything is its pure will and knowing, as Hegel states: "the objective element into which it puts itself forth, when it acts, is nothing other than the self's pure knowledge of itself" (PoS §792, 481, 482). Pure will demands pure laws. Rigid moral systems emerge, offering no room for error. But as moral consciousness demands action, and as all consequences of actions are unpredictable, "errors of ignorance" cannot be sufficiently addressed (Harris 1997, 720). To attain a more accurate understanding of human beings, we therefore require the inclusion of forgiveness (see Harris 1997, 720). While the ideal moral consciousness says 'no,' the forgiving community recognizes the practical circumstances of action and says 'yes' (see Harris 1997, 720).

The unification through forgiveness brings self-consciousness to a preliminary identity expressed in the abstract formula of 'I'='I'. This formula represents the reconciliation within a forgiving society, where everyone recognizes him or herself in each other and on this basis can forgive. However, as forgiveness is primarily a result of an established practice, the formula falls short of entirely capturing the essence of self-consciousness. The formal idea of Spirit's existence is not enough (see PoS §793, 482). Spirit has to be recognized in its historical realization.[51]

To introduce the solution of the so-called Concept, Hegel first analyzes the problem in more detail. According to Hegel, the two extremes that have to be reconciled are:

a Self-consciousness which "flees from contact with the actual world" (PoS §658, 400). Hegel describes this as a "self-willed impotence to renounce its self" (PoS §658, 400). So this self-consciousness stands in "absolute difference" to the experiencing, acting self (PoS §658, 400). "Devoid of substance," this consciousness finds itself as an unhappy, lost, yet "beautiful soul" (§658, 400, see also PoS §795, 483). We can also call this 'empty self-consciousness' or 'spirit for-itself'.

b The opposite of the beautiful soul is "its realization", the religious act, so that the beautiful soul disappears in "thin air" (PoS §795,483). It is the religious community that obtains spirit as it is in-itself.

Now,[52] Hegel expounds on the unification of the two sides in 'Absolute Knowing':

51 Rockmore writes on this: "It is obviously not sufficient to have an idea. The idea must also be realized" (Rockmore 1997, 174).

52 I skip here Hegel's remarks on religion as they are not as important for the development of my arguments of Hegel's historicity. Nevertheless, for a complete understanding of the chapter, we have to take this historical becoming of Spirit through religion seriously. Harris, for example, remarks: "The lesson that philosophy is not to be understood apart from its history is widely understood; what Hegel's science of experience teaches us is the much more demanding imperative that philosophy and religion must be comprehended together in the context of the actual history of the human community" (1997, 721).

> The unification of the two sides has not yet been exhibited; it is this that closes the series of the shapes of Spirit, for in it Spirit attains to a knowledge of itself not only as it is in itself or as possessing an absolute content, nor only as it is for itself as a form devoid of content, or as the aspect of self-consciousness, but as it is both in essence and in actuality, or in and for itself.
>
> PoS §794, 483

Hegel's solution can be summarized as follows: Self-consciousness must, through reflection, understand its manifestation in religious consciousness independently, without relying on external factors. By acquiring historical knowledge, self-consciousness should transform into a self-aware absolute unity. This transformation allows it to interpret all past manifestations of Spirit, enabling consciousness to recognize its own existence within Spirit. Hegel's approach involves comprehending this historical journey, reaching its culmination in religious spirit, which then transitions from religion's intuition to philosophy's rationality. In the historical stage of philosophy, the Concept fulfills the vital task of uniting both ends of consciousness and self-consciousness, as Hegel elucidates:

> it is the Notion [the Concept] in its truth, viz. in unity with its externalization; it is the knowing of pure knowledge, not as an abstract essence such as duty is, but of knowledge as an essential being which is this knowledge, this pure self-consciousness which is, therefore, at the same time a genuine object, for the Notion is the Self that is for itself.
>
> PoS §795, 485

Hegel introduces us to the criterion for Absolute Knowing, which involves comprehending all of Spirit's moments through the singular concept known as 'The Concept'. This concept is simultaneously subjective and objective. It is generated subjectively by individuals and simultaneously provided by the community, bridging, for example, the gap between personal will and communal norms. The Concept is powerful enough to serve as the blueprint for how

---

Harris highlights further that the merely formal unity cannot be understood apart from its historical becoming in religion. The arrival of the Christian savior and his demand for forgiveness is a practice through which religion precedes our conceptualization. In other words, the formal unity is not the ahistorical end of Hegel's endeavor. The whole project of Absolute Knowing is rooted in its historical becoming, a point that Brandom arguably does not touch in deeper detail.

human societies have evolved throughout history. It stands as the objective criterion through which individuals can understand themselves without external dependencies and also describes the many processes that have guided the development of societies towards rationality, towards the Concept. Now, our task is to elaborate on The Concept as the solution and demonstrate its practical presence in historical contexts.

4.3 *The Unification of Consciousness with Self-Consciousness through the Concept*

From the former exposition of the problem, we know that the unification of the extremes of consciousness has to be grasped through the unity of the Concept (PoS §796, 484). In particular, it is the "simple Notion [Concept] which has surrendered its eternal *essence*" (PoS §796, 484). Though the simple Concept is not the inwardness of the beautiful soul, it is also not the externalization of the acting consciousness. It is the sublation of both. The advanced Concept can thereby be powerful enough to incorporate the contradiction between both sides—the beautiful soul and the acting consciousness. Instead of discussing the simple Concept as a formal solution, however, Hegel follows here what I call his phenomenological approach. His goal is to observe how humans act in society in order to grasp their spiritual becoming representationally, and then to eliminate the representational moments within our understanding of it. As a result, we will arrive at a more advanced Concept. We can call this a historical-phenomenological reduction. Ultimately, we do not show that humans are concepts—as it is often assumed—but that in their historical moment, they can be objectively understood through their *conceptual* activity. In this regard, Rockmore summarizes the importance of the subject's conceptual activity:

> For philosophy, the importance of consciously acting is that the subject 'carries out the life of the absolute spirit' (§776, 484*). As concerns the concept, the final step is that 'what happens in principle is at the same time explicitly for consciousness' (§796, 484) or actually carried out. Through human action, the contradiction between what is the case in principle and actuality is overcome.
>
> ROCKMORE 1997, 187–188

The main idea is that the Concept is the "self's own act" (PoS §797, 485), which occurs through the interactions of humans who create a common conceptual sphere, i.e., a spirit. The subjects carry out "the life of the absolute spirit" (PoS §796, 484) in a conceptual activity. Absolute Spirit occurs therefore first during the historical stages of its communities in an implicit form—the conceptual activity—and comes through more and more reflective communities to its self-knowing—The Concept. This is because the life of the community is grounded on self-consciousness' self-renouncing act, i.e., action. In other words, it relinquishes its formal, beautiful, and moral unity, leading to actions that give rise to mistakes, and in doing so, it renounces its former self. The renouncing act is repeated by each individual consciousness in the community through their individual action. Each individual action, however, is guided through concepts because without having a concept guiding one's actions, it is implausible to say that one acts at all. Because of the diversity of individual actions and conflicts, communities need to build higher and higher plateaus of conceptual self-recognition. We see this, for example, in the increasing development of more and more complex legal systems with increasingly complex concepts. By introducing more and more concepts utilized in institutions, human communities understand and guide more and more their own doing.

We get a further glimpse of the works of the Concept by looking at the division between good and evil. According to the former analysis, good and evil are dialectically related to each other as not acting (in the form of the beautiful soul) and acting. The "self-sundering, or the negativity which the Notion is" (PoS §797, 485) is evil. In other words, the Concept releases itself in action and thus damages its ideal unity of the beautiful soul. This is because actions inevitably lead to unforeseeable negative consequences. Consciousness as this conceptual activity becomes its "own act of renunciation, just as the Notion which it renounces is its own Notion" (PoS §796, 484). For Hegel, the structure of the Concept is appropriate to describe the "determinateness of both" (§796, 484) sides of consciousness. It is an ideal unity that mediates itself through action, that grounds the moral evaluation of this action, and that 'forgives' through the renouncing of its own renouncing.

Hegel emphasizes that action is a negative conceptual activity. He thereby relates actions to concepts. Now, self-consciousness can slowly approach the activity of the Concept. It arrives at the "unity of knowing" by neither understanding society intuitively nor representationally, but rather through the action of structuring societies conceptually (PoS §796, 484).

The Concept is the fundamental tool of Hegel's philosophy to describe the formation of societies towards their active self-recognition. My task is now to show how it is, in fact, historical.

4.4 *Absolute Knowing and Human History*

Before exploring the dialectics of the final conceptual form of Spirit and its role in history, let us summarize the discussion thus far. Initially, knowledge exists abstractly within consciousness, as it apprehends something distinct from itself (see Fulda 2007, 350). The desire for a complete understanding of the object in-itself that only appears for-itself drives consciousness beyond itself (see Fulda 2007, 350) resulting in various self-consciousnesses. Through several sublations, these diverse forms eventually find unity in religious Spirit. However, religious Spirit also consists of two contradictory parts. An acting consciousness stands in an extreme tension to an inner beautiful soul of consciousness (see Fulda 2007, 350). While communities externally reconcile the tension of ideal and action through the practice of religion, true reconciliation has yet to be achieved. To accomplish this, religion must be guided by a concept accessible to all rational subjects rooted in their original, negative activity. It is the negative activity that is the ground for consciousness and for society. Fulda makes the important point here that the substance of this religious community has to become subjective (see Fulda 2007, 350–351). It means the historical state of religion has to be rationally understood. Here, Hegel combines two ideas to achieve 'Absolute Knowledge' of this substance. Firstly, he introduces the idea of the Concept, which is rich enough to express the religious practice beyond mere intuition. Secondly, he introduces the idea that human beings have always already employed this concept to construct their community within their historical moment. Through exploring the relationship between the self-knowledge of the Concept and its historical generation and utilization in society, we reach the "last shape of Spirit" that manifests its complete and true content as the Self (PoS §798, 485).

To elucidate the historical development of the concept leading to self-awareness, Hegel steers us from the initial formal notion of the Concept towards its evolution in human history. We begin by reiterating the subjectivity of the 'I,' emphasizing that its comprehension is intrinsically linked to its historical milieu. The 'I' exhibits a unique duality as both an entity unto itself and simultaneously as part of a mediated universal 'I. Regarding this, Hegel asserts: "it is 'I', that is this and no other 'I', and which is no less immediately a mediated or superseded universal 'I'" (PoS §799, 486). Hegel then delves into the intricate relationship between the universal and individual facets of the 'I,' acknowledging that a comprehensive understanding, given its inherent self-differentiation, can only be attained within a community. In reference to this self-differentiation, he articulates:

> It [the 'I'] has a content which it differentiates from itself; for it is pure negativity or the dividing of itself, it is consciousness. This content is, in its difference, itself the 'I', for it is the movement of superseding itself, or the same pure negativity that the 'I' is. In it, as differentiated, the 'I'

> is reflected into itself; it is only when the 'I' communes with itself in its otherness that the content is comprehended.
>
> HEGEL, PoS §799, 486

In essence, it is the self-differentiating nature of the 'I' that allows for the potential realization of the unity between the universal and individual aspects. However, Hegel emphasizes again that the problematic relationship between the universal and individual cannot be resolved through a mere abstract formula such as 'I'='I'. According to Hegel, such a merely subjective resolution alone is insufficient. Instead, the intricate relationship between the 'I,' 'you,' and 'we'—the community of 'I's—must be understood within the framework of an actual community where everyone operates as an individual 'I' and treats one another as an 'I'. As discussed earlier, this transition marks Hegel's exploration into the realm of Spirit:

> Stated more specifically, this content is nothing else than the very movement just spoken of; for the content is Spirit that traverses its own self and does so for itself as Spirit by the fact that it has the 'shape' of the Notion in its objectivity.
>
> PoS §799, 486

It is through Spirit, rather than the simplistic equation of 'I'='I', that the true resolution to the ontological difference can be achieved. However, in order to grasp this essence of Spirit, we must now reconstruct our understanding of the connection between the manifestation of Spirit within a historical context and the actions of historical subjects who are guided by the Concept.

It is of utmost importance to highlight that our pursuit of Absolute Knowing is intricately intertwined with an existing Concept that operates consistently throughout history. Consequently, it is not merely a formal, external criterion for science but rather a criterion that is internally derived from its historical manifestation within society.[53] However, the realization of 'Absolute Spirit' is

---

53 Rockmore highlights that the attainment of the resolution leads humans to the standpoint of science (1997, 189). By transcending the ontological distinction between subject and external object, a moment of self-certainty arises, where Hegel declares that the "problem of knowledge [...] comes to an end" (Rockmore 1997, 188). According to Hegel, attaining the standpoint of science necessitates demonstrating the presence of Spirit as a concept within consciousness: "Spirit, manifested or appearing in consciousness in this realm, or more precisely, produced in it by consciousness, is Science" (Hegel, PoS §798, 486).

contingent upon "Spirit […] knowing what it is" (PoS §800, 486), making it an ideal goal that remains yet to be fully attained. Nevertheless, since consciousness constantly strives to transcend its limitations and moves towards absoluteness, this drive is also present within society. Given these circumstances, Hegel employs a specific method, which I referred to as a form of phenomenological reduction before. Within this framework, the aim is to eliminate all external forms of understanding within the historically existing Spirit. By doing so, the underlying mechanisms of the Concept can be uncovered and Absolute Knowing can be established.

To accomplish this 'phenomenological reduction', Hegel must address the challenge concerning the temporal relationship between religious spirit and the self-awareness of spirit. The predicament lies in the fact that our understanding of "substance […] exists earlier than […] its Notion-determined 'shape'" (PoS §801, 486). This is because "substance […] as yet undeveloped in-itself or the Ground and Notion in its still unmoved simplicity […] the inwardness or the Self of the Spirit" (PoS §801, 486) has not undergone full development; it exists only in representation. In other words, substance lacks a self-identity. Consequently, we have encountered substance solely as the abstract (refer to PoS §801, 486), historical moment of religion. At this juncture in our endeavor to comprehend Spirit, we encounter the complex concept of the elimination or annulment of time, which has posed challenges for many interpreters and has been a source of confusion in the literature. For instance, Habermas, as discussed earlier, interprets these passages as evidence that Hegel embraces ahistoricism.

The misunderstandings of Absolute Knowing are closely tied to the paradox elucidated by Hegel, which states that the whole of the concept within consciousness precedes its temporal moments:

> In the Notion that knows itself as Notion, the *moments* thus appear earlier than the *filled* [or *fulfilled*] whole whose coming-to-be is the movement of those moments. In *consciousness*, on the other hand, the whole, though uncomprehended, is prior to the moments
>
> HEGEL PoS §801, 487

Can Hegel argue that Absolute Knowing, which involves understanding consciousness' own activity, surpasses the limitations of time? Absolute Knowing is not completely detached from time. Rather, it both precedes and is influenced by temporal moments. We have to investigate this problem in more detail in the subsequent points.

4.5 *The Contradiction of Time and a Science of Experience*

We have reached the ultimate paradox in our quest to attain Absolute Knowing. On the one hand, concerning the self-knowing Concept, its moments of self-knowing precede the whole (see PoS §801, 487). On the other hand, the non-conceptualized [das Unbegriffene] whole emerges prior to the moments in intuition (see PoS §801, 487). Moreover, the understanding of this whole through the concept of time remains inadequately conceptualized, as Hegel describes it, for instance, as empty: "Time is the Notion itself that is present and appears to consciousness as empty intuition" (PoS §801, 487). Stripped of subjectivity, the concept of time situates historical events on an infinite time-line, devoid of conceptual significance. It presents historical events as a contingent succession of events, bound together solely by their temporal sequence. Substance, when viewed through such a concept of time, becomes entangled in an incessant motion, lacking any sense of conceptual coherence. In this context, Hegel advances his critique of the intuition of time and emphasizes the need to set it aside. He states, "It is the outer, intuited pure Self which is not grasped by the Self, the merely intuited Notion; when this latter grasps itself it sets aside its Time-form, comprehends this intuiting, and is a comprehended and comprehending intuiting" (PoS §801, 487). In essence, Hegel asserts that the external aspect of the Self, perceived through intuition alone, cannot be fully understood by the Self. The Self must transcend reliance on intuition and grasp its own essence, relinquishing the constraints of time in order to attain a comprehensive and reflective understanding. The nature of experienced substance relies more on conceptual understanding rather than its mere integration within a timeline. From my perspective, Hegel aims to develop a comprehensive understanding of history by considering the perspective of finite human communities in their specific historical moments utilizing concepts. The accurate concept of this historical substance encompasses experience rather than time, as Hegel states:

> For this reason it must be said that nothing is known that is not in experience, or, as it is also expressed, that is not felt to be true, not given as an inwardly revealed eternal verity, as something sacred that is believed, or whatever other expressions have been used. For experience is just this, that the content—which is Spirit—is in-itself substance, and therefore an object of consciousness.
>
> PoS §802, 487

Hegel argues that the content of experience is substance itself. As the entirety of substance is initially intuited before its subjective conceptualization, it is acquired through experiential encounters with represented objects. While it is possible to perceive these objects in relation to time, such categorization represents a simplistic classification. For instance, stating that individual *A* had a car accident in 2018 and died 3 days later fails to establish a coherent connection between these events. Utilizing the concept of time, Spirit, in particular, cannot be adequately understood with regard to its inherent conceptual structure, content, and rational progression either. Our attention turns towards the development of a scientific framework that can provide conceptual clarity to these experiential events related to Spirit. Now, the emphasis lies more on conceptual relationships rather than temporal sequence and intuition. Consider, for instance, the cyclical nature of plants. In conceptual terms, the question of whether the bud precedes the blossom or vice versa becomes less relevant—more importantly, they are interconnected moments within the nature of plants overall. In essence, the conceptual hierarchy diverges from the linear progression of time.

As science begins and ends with experience, the unity we seek is not based on an assumption of an external entity or intuition, but rather on an understanding of what happens during experience. Hegel, therefore, defines his project also as circular: "The movement is the circle that returns into itself, the circle that presupposes its beginning and reaches it only at the end" (PoS §802, 488). The circle has no strict before and after. Moreover, the circularity also implies that substance never existed external to us in time; rather, we merely conceptualized it in a mistaken manner. Now, however, we engage in the subjective experience of substance and reflect upon its becoming, leading us to the realization that substance and subject are identical in The Concept. The presence of the Absolute permeates our existence as experiencing and historical human beings, who utilize concepts. In the conceptual order, we sublate the moment of time and intuition.

Fulda provides additional insights into Hegel's concept of the Absolute in relation to experience. He specifically highlights that substance serves as the object of consciousness (see Fulda 2007, 375). However, the reflection on this object does not commence with something external or purely internal; rather, it originates from the experiential aspect of the identity between substance and subject (see Fulda 2007, 375). Hegel proposes this resolution of science's inception in substance and subject in §804. Fulda emphasizes that while this beginning remains a historical moment, it is one in which we have no compelling reason to surpass, as we have achieved identity. This Absolute Knowing remains valid unless new contrary experiences emerge:

> Its knowledge corresponds to its standard, and its 'concept to the object,' as well as 'the object to the concept' […] at least insofar as, at present, no experiences are formed in consciousness that arise from any moment of substance already incorporated into comprehension or yet to be incorporated, according to which, the absolute knowledge (that is already attained and not limited by anything external to it and not blurred by anything internally foreign) does not have to be corrected. However, it is not claimed that this will remain the case in the future.
>
> FULDA 2007, 377, translation by me[54]

In alignment with my perspective, Fulda recognizes that Absolute Knowing does not preclude the possibility of subsequent impactful experiences that can reshape our understanding. Consequently, Hegel's philosophy does not represent a conclusive or ultimate standpoint. Instead, it commences with experience, comprehends the manner in which experience unfolds, attains identity with its experiences, and maintains openness to further historical experience.

At this juncture, we encounter another instance of the divergence between Brandom's interpretation and my own understanding of Hegel's philosophy. In my view, Hegel presents a science of experience that is rooted in the historical development of reflective human beings who contemplate their past experiences through historical concepts. The resulting scientific framework consisting of concepts is subject to potential evolution and transformation over time and only for now arrives at The Concept. On the contrary, Brandom advocates for conceptual realism, positing that rational subjects engage in language games. These language games are guided by the notion of an ahistorical, independent conceptual reality, while the subjects' actions strive to align with it. According to Brandom's perspective, the resulting scientific understanding remains stable and is not prone to radical changes in response to the emergence of new experiences. He arrives at the, in my mind, flawed conclusion that the structure of mind-external reality must align with that of mind-internal

54 "Sein Wissen entspricht seinem Maßstab und sein 'Begriff dem Gegenstande' sowie 'der Gegenstand dem Begriffe' […] zumindest insofern, als sich gegenwärtig von keinem schon ins Begreifen aufgenommenen oder neu aufzunehmenden Moment der Substanz aus im Bewußtsein Erfahrungen bilden, gemäß denen das durch jenes Erkennen erreichte absolute Wissen mit seinem von nichts ihm Äußerlichen mehr eingeschränkten und von keinem ihm innerlich Fremden mehr verwischten, allumfassenden Umfang noch korrigiert werden muß. Daß das auch in Zukunft so bleiben wird, ist damit nicht behauptet"(Fulda 2007, 377).

concepts. This foundational understanding sets the stage for future endeavors aimed at approximating mind-external reality.

4.6 *Externalizations of Spirit—Time and History*

Having already elucidated the fundamental contours of Absolute Knowing and its intricate connection to history and experience, our focus now shifts to the interrelationship between time and history.

Time is the abstract understanding of the externalization of Spirit in nature (see Hoffmeyer 1992, 198). As such we intuit events on a timeline. History, however, is the externalization of nature through spirit (see Hoffmeyer 1992, 198). In simpler language, time is a phenomenon that belongs to the concept of nature, while history is a social phenomenon that belongs to the concept of Spirit. We can reconstruct these ideas and their relation to each other in the following sections.

4.6.1 Definition of Time

In the *Philosophy of Nature,* Hegel demonstrates the relation between time and space as time deduced from space. Space is the result of the logical idea, spirit in-itself, freely releasing and externalizing itself as nature. The first externalization of nature is therefore space in its three dimensions. The second externalization of this externalization is time. Since this concept of time, however, is only abstract, Hegel further derives movement and motion in pushing the concept of nature further and further beyond itself. In a word, nature is neither simply space, nor time. In this sense, time is only the concept of our naturalist attitude when we look at an externalized world. Time is the concept of intuition. It can deliver a preliminary foundation for understanding further concepts of nature. Nature itself, however, has to be grasped in a higher concept since it is neither space nor time.

As concerns time, I agree with Hoffmeyer: "when the concept grasps itself as itself, it 'sublates its time-form'" (1992, 200). So we can overcome the intuition of time through the concept. Time therefore is only an abstraction of Spirit's movement that leads to contradiction and pushes then toward the overcoming of itself. It is, however, true that Spirit can relate to itself in the intuition of time. So, for example, we see different historical principles contingently following each other in world history. Through time, we intuit such historical movements and express them then in principles. Movements can, therefore, occur in the abstract notion of nature, or in the notion of World Spirit. In any case, however, time is related to intuition, and as Hegel argues, at the beginning of his section on 'Absolute Knowing', for achieving an actual understanding of ourselves, intuition has to be sublated.

Hoffmeyer writes about the overcoming of time through Spirit: "[The movement of Spirit] leaves behind time as it is understood in abstraction from the movement of Spirit" (Hoffmeyer 1992, 200). So time is left behind. Even though we could still look at history in terms of time, this view would constitute only a contingent becoming of Spirit, as Hegel writes, "in the form of free contingent happening" (PoS §807, 492). It would constitute world history without understanding how it takes place. We would recognize historical events as a ceaseless succession of events on a timeline, yet not understand why they occur. So, in a sense, the time of Spirit can be intuited, but our goal is to conceptually grasp Spirit's movement. As an ultimate principle of unity, Absolute Knowing orders all occurrences of our experience coherently, not only with regard to time, which would be an external, intuitive solution but with regard to their necessity. In other words, Absolute Knowing, following Fulda's thesis, serves as a principle for understanding World Spirit organically.

In *Vorlesungen zur Philosophie der Weltgeschichte* [Lectures about the Philosophy of World History], Hegel distinguishes 'original history' that captures the first occurrence of spirit in time, 'reflected history' that already grasps different forms of world history, and a 'philosophical world history' (see Hegel, 2015, 3)[55] that orders all world-historical occurrences with regard to one principle. 'Philosophical world history', however, is not based on an arbitrary universal [Allgemeines], but rather on a concrete universal [konkretes Allgemeines]. It is the spiritual principle of the people and the leading soul as Hegel explains (see Hegel, 2015, 14). It has to be captured as its own activity. It is the historical production of its own occurrence. For Hegel, it therefore achieves a form of identity: "Because Spirit is always with itself" (Hegel, 2015, 15).[56] It is Spirit that can remain its self-producing identity for all futures to come, but it is not a mind-external eternity.

It is, of course, questionable whether Hegel could really deliver such a unifying principle in an explicit shape. However, one concept which clearly fails as such a principle is history as a progression in time. Without appropriate, conceptual understanding, but only intuited through time, spirit would be "pure restlessness, total absence of stability [...] neither motion nor time" (Hoffmeyer 1992, 200). So, the substantial principle of order that we attempt to grasp must be conceptual. Substance cannot be intuited as a movement in

55 Hegel, G.W.F. "Vorlesungen über die Philosophie der Weltgeschichte. Band 1: Die Vernunft in der Geschichte." *Gesammelte Werke* 27. Edited by Bernadette Collenberg-Plotnikov. Hamburg: Meiner, 2015.

56 "Denn der Geist ist ewig bei sich selbst"(Hegel, 2015, 15).

time, a ceaseless becoming. We attempt to point out the principle as a conceptual resting point, in which spirit can realize how it became and remains what it is.

I have interpreted Hegel's philosophy as a relativist historicism. Although I believe that this is what he suggests, whether the approach itself is successful is a different question. Habermas, however, dismisses Hegel's account because he understands Absolute Spirit as ahistorical. As it should be clear by now, the principle of Absolute Knowing is intended to be timeless, yet, explicitly historical. Nevertheless, even if it is intended to be historical, can we really eliminate all external elements[57] in reflection and only relate to our own historical becoming, as Hegel proposes? Given the complexity of history, this task of reflection remains a challenge for each historical moment and it does not end with Hegel's *Phenomenology of Spirit*.

4.6.2 Definition of Nature and History

Nature, which is the externalization of the logical Idea, expresses the immediate becoming of Spirit. In its immediacy, however, it remains abstract. History, on the contrary, is a self-relation of Spirit. We could say that as the reflected philosophy of history it is the project of coming to know what it really is. Hoffmeyer summarizes both concepts: "Nature is Spirit's immediate process of coming to be itself. History is the self-mediating process of Spirit coming to be itself" (Hoffmeyer 1992, 198). To say it simpler, Hegel distinguishes nature and history as two forms of how spirit becomes. He writes on nature as resulting from Spirit in the *Phenomenology*:

> This last becoming of Spirit, Nature, is its living immediate Becoming; Nature, the externalized Spirit, is in its existence nothing but this eternal externalization of its continuing existence and the movement which reinstates the Subject.
>
> HEGEL PoS §807, 492

Nature can be seen as the Spirit that manifests itself to us solely in its external form. I understand this concept by observing phenomena in daily life (like tables and chairs), perceiving them as self-existing entities separate from myself. I represent them and rely on the notions of time and space to comprehend their location. However, a phenomenological examination of the object

57 Hegel's account is rooted in the elimination of all presupposed ideas and thoughts when contemplating Absolute Spirit. This process ensures that it becomes a unit of thought uncontaminated by anything external to itself.

would uncover that it is not solely confined to an immediate, infinite external realm, but that my subjective activity plays a role in constructing it. This brings us back to the question of self-consciousness, concepts and their relation to Spirit.

In the *Philosophy of Nature*, Hegel aims to demonstrate the process by which nature is transformed into Spirit through the revelation of its inherent contradictions. As our understanding of space and time merely provides us with a superficial depiction of objects, Hegel believes that understanding history brings us to the conceptual side of Spirit. This idea is already present in his *Phenomenology:*

> But the other side of its Becoming, History, is a conscious, self-mediating process-Spirit emptied out into Time; but this externalization, this kenosis, is equally an externalization of itself; the negative is the negative of itself. This Becoming presents a slow-moving succession of Spirits, a gallery of images, each of which , endowed with all the riches of Spirit, moves thus slowly just because the Self has to penetrate and digest this entire wealth of its substance.
>
> HEGEL, PoS §808, 492

In this passage, Hegel explores the concept of history. He posits that history is not simply a linear progression of moments in time, but rather a self-negating process that gives rise to a sequence of Spirits. While time is tied to immediate sensation and lacks the ability to capture conceptual movement,[58] history represents the conceptual movement of Spirit. This distinction arises from the fact that history is inherently linked to consciousness and knowing, as it involves the activity of conceptualization. The diverse manifestations of Spirit are subjectively mediated by self-aware individuals who employ concepts in this process.

Furthermore, this process must be connected to the objects that shape the history of individuals. These individual subjects are not abstract or immediate entities but rather active participants in a history of experience. According to Hegel's theory, these self-conscious, experiencing individuals cannot exist without consciousness of something, and this consciousness entails a tangible process of mediation. Therefore, history is not merely a movement of

58 Conceptual movement transcends notions of before and after. For instance, consider the relationship between the roots and leaves of a tree. It becomes challenging to determine whether the roots cause the leaves, or the leaves cause the roots. Both elements follow a conceptual order, and neither can be cogitated as prior to the other in time.

impersonal substance but rather a movement of a 'subjective substance', i.e., subjects actively making decisions on the basis of their experienced objects.

Hegel establishes a close link between the progression of history and the development of individuals attaining higher levels of self-consciousness, which he refers to as 'Bildung'. This concept revolves around the formation and deepening understanding of individual consciousnesses. Absolute Knowledge, which encompasses the relationship between consciousness within a specific historical moment and self-consciousness as a self-relation within that moment, is dependent on the historical context. This knowledge cannot be explained solely through the intuition of time as a mere succession of stages, where Spirits would appear as entities causing one another. Instead, Spirits are bound by the actions of individuals guided by socially mediated principles and their intuitive comprehension of them. Thus, it becomes challenging, if not impossible, to isolate a single historical event as the cause for another event. This perspective on causality would be sociologically simplistic. Absolute Knowing, through reflection on preceding Spirits, achieves independence from time and natural causality. The ultimate goal of these movements is the realization of Absolute Spirit, which connects these stages based on conceptual ideas. It involves a process of social reflection, whereby Spirit investigates the diverse historical principles that guide societies at different times, enabling self-discovery within the historical process that transcends mere causality or natural determinism.

4.6.3 Hegel's Solution of Absolute Knowing

To reiterate, according to Habermas, it appears that Hegel not only annuls the concept of time as a contingent form of intuition but also rejects historicity, thereby advocating for an ahistorical approach. However, in light of Hegel's differentiation between time as an intuition and history as a conceptual movement of Spirit, this interpretation should now be seen as incorrect. Absolute Knowing is inseparable from its historical context and arises from the tangible historical progression of ideas involving active individuals. In this final point, I will examine Absolute Knowing as the notion that knowledge claims are confined to the historical moment.

To present the concept of historicity convincingly, let us begin by assuming that human understanding exists outside the confines of the movement of consciousness and its historical contexts. I believe that this leads to a form of empiricism that only offers a partial understanding of experienced objects based on immediate sensations. There are many challenges for such a raw empiricism. Instead, Hegel suggests a systematic, historical approach to experience as the basis for natural consciousness, wherein immediate sensations are reconstructed as a systematic part of a larger process. To comprehend this process, individuals themselves must recognize the interplay between consciousness and the perceived object. We can refer to this process of knowing as 'history', depicting it as the unfolding narrative of how consciousness establishes connections with objects. For consciousness to attain knowledge, it must also comprehend itself. In the case of self-consciousness, its own historical generation becomes crucial. Our personal development occurs within society, wherein different roles are assigned to individuals, influencing their self-awareness in response to societal expectations. Diverse perspectives emerge concerning political institutions and artifacts, as well as religious aspects that serve to unite our society. Additionally, varying viewpoints on science further shape our interactions, either fostering unity or fostering diversity. These perspectives are rooted in active communities with distinct histories. Our individual actions result from a series of mediations within historical society, while they themselves integrate experiences into a communal body of knowledge. This communal body of knowledge forms a culture that is a result of its historical moment and substantiates this historical moment. Ultimately, these collective, historical bodies—Hegel calls them Spirits—occur in an observable world history, which we study and use as a basis for understanding ourselves.

Hegel explains the emergence of histories through the concept of Absolute Knowing. By this viewpoint, these otherwise contingent events constitute a series of necessary events for the formation of societies recognized as the history of their present moment. Absolute Knowing can then be understood as a criterion for a unified history, namely our own history, by which, moreover, we can conduct our pursuit of knowledge. Most importantly, Absolute Knowing elucidates this unified history without relying on non-conceptualized arguments and does not seek external explanations for how we arrived at our historical moment. In Absolute Knowing our focus lies in examining our position within our historical context. Hegel does not venture beyond this historical standpoint.

In essence, Hegel offers a historicist solution to the problem of knowing and truth by asserting that the spirit present in all these subjective actions throughout history cannot simply be an unorganized substance that randomly manifests in world history. Likewise, it cannot be a subject that imposes unexamined methods in order to comprehend the objects of history. Rather, it is the subjective essence that arises from its own historical process and achieves self-clarification. Hegel articulates this idea of a subjective substance by stating: "Spirit, however, has shown itself to us to be neither merely the withdrawal of self-consciousness into its pure inwardness, nor the mere submergence of self-consciousness into substance" (PoS §804, 490). Subjectivity and substance are captured in one movement of spirit. He proceeds, to explain this relationship:

> this movement of the Self which empties itself of itself and sinks itself into its substance, and also, as Subject, has gone out of that substance into itself, making the substance into an object and a content at the same time as it cancels this difference between objectivity and content.
>
> PoS §804, 490

Hegel identifies the process of the movement between subject and substance as "reflection out of immediacy," signifying the subject's ability to step back and observe itself. He describes this as "the Notion's separation of itself from itself" (PoS §804, 490). However, this "pure act of 'I' = 'I'" represents an indeterminateness that gives rise to another movement (PoS §804, 490). Hegel attempts to capture this movement towards the conceptualization of its own principle through the Concept. He disqualifies a possible resting position of this Concept in abstraction "as if it were afraid of the externalization of itself" (PoS §804, 490). Instead, Spirit as the social unity of the 'I's in world history has to demonstrate its identity "in its externalization" (PoS §804, 490).

Hegel highlights the importance of considering self-awareness as a procedural unity of negation. This suggests that Spirit is in a constant state of becoming through the process of negation. It is argued that a mere introspective examination of ourselves as objects is insufficient for this understanding. According to Hegel, our capacity to introspect resembles the nature of history itself. It relies on self-reflection, which necessitates both externalization and the subsequent negation of this externalization. Consequently, the Concept is not an abstract transformation of, for instance, a substance existing in abstract time and space. Rather, the Concept manifests itself within a concrete historical context by negating previous forms of knowledge. Through observing these negations, we come to realize that no moment remains stagnant in itself and that each moment contains inherent contradictions. Collectively, we learn from diverse historical examples that this movement forms circular patterns of contradictions and reconciliations. The resulting super-circle is not an abstract notion within Absolute Knowing but rather represents our history. The conceptual differentiation of our history is synonymous with the circular differentiation of ourselves. In this sense, Hegel's philosophy retains an empirical character since Absolute Knowing is derived as an integral part of our world history and historical comprehension. To be more precise, Absolute Knowing marks the culmination of a world-historical era where the Spirit grasps its own essence through the Concept. It is one circle of reflection that comes to an end.

Hegel explains that the Concept is "the movement of carrying forward the form of its self-knowledge [...] the labor which it accomplishes as actual History" (PoS §803, 488). Still, we have to understand the Concept through an appropriate understanding of history. In §803, Hegel proposes therefore a very brief history of philosophy that moves from the first insights of religion to philosophy and The Concept. First, we discover the identity of Thought and being. Second, the identity becomes substance. Third, we express this as the identity of being and externality. Fourth, this leads to the insight 'I'='I'. Since, fifth, this substance is in movement, we have to understand it as an identity of thought and time. Yet, there is still an external element left, as Hegel writes: "But the difference left to itself, unresting and unhalting Time, collapses rather within itself; it is the objective repose of *extension*" (PoS §803, 489). Hegel comes to the conclusion that the identity of Being and extension, and the identity of Being and time are not yet reconciled. Substance is not yet subject. Hegel completes this very brief historical summary of philosophy's development in §804. There, he offers his solution of a subjective substance and a substantial subject. In §805, Hegel explains the only meaningful solution to the historically remaining problem of the ontological difference. Spirit has "shaped itself [...] burdened with the difference of consciousness" (PoS §805, 490). It has clarified itself by means of negations that led to reflections of what it is for itself. Finally, Hegel notes that in this historical process, we have "won the pure element of its existence, the Notion" (PoS §805, 490). The Concept is the final result and is demonstrated as the core of how we can understand ourselves. Hegel finishes his argument for the Concept here.

The Concept is the central concept as demonstrated through the history of philosophy. The content of the concept is difference in-itself. In this difference, it knows how it alienates itself (see PoS §805, 490–491). Alienation introduces the need for the content (see PoS §805, 490–491), while it is a "restless process of superseding itself, or negativity" (PoS §805, 491). Content never remains in-itself, but reveals itself as identical with its becoming, since the "content is the Notion" (PoS §805, 491). Spirit thus inhabits both "its existence and movement" by virtue of the fact that this "is Science" (PoS §805, 491). All differences of consciousness are therefore unified in the "Notions [...] as their organic self-grounded movement" (PoS §805, 491).

Science is the one Concept that explains, according to one category, namely the Concept, how it contains difference, how it cancels, and reconciles it. Hegel's criterion of truth is met here, since the object as Truth and the subject as knowing come together in a single "immediate unity" that is the Concept (PoS §805, 491). There is no "passing back and forth" since the Concept can be seen as the working force behind all of its occurrences (PoS §805, 491). As the

Concept has been the content of Absolute Spirit all the way through history, and as it grasps its form, we can call this Absolute Knowing. Hegel summarizes the last step of his project:

> To know the pure Notions of Science in this form of shapes of consciousness constitutes the side of their reality, in accordance with which their essence, the Notion, which is posited in them in its simple mediation as thinking, breaks asunder the moments of this mediation and exhibits itself in accordance with the inner antithesis.
>
> PoS §805, 491

The result is the Concept that knows itself and even includes its own antithesis. However, with this last move, the Concept as the substance of former history is not yet completed with respect to further movements. Hegel asserts that the limit of the concept is its beyond. From this, I infer that history has not reached its conclusion but continues to proceed. Nevertheless, we now possess the ability to scientifically understand and approach our ongoing transformation.

In my view, it is evident that the Concept serves as a working definition. It is the most effective epistemological strategy to elucidate our history of knowledge. The Concept can now be applied to all aspects of existence. However, its effectiveness must be demonstrated through its application and verification [Bewährung]. Thus, it is inaccurate to claim that the Concept enables us to predict the complete structure of 'the' system. Rather, it provides us with an understanding of what it means to engage in scientific inquiry and work systematically.

In his final argument, Hegel provides clarity on why the Concept not only emerges within history but also remains confined to it. Hegel's explanation unfolds from the internal antithesis inherent to the Concept of the formation of an externalized nature. While we cannot surpass the attained standpoint of science through mere reflection, and even though we have reached the culmination of our current epistemological investigation, we can still explore the potency of the Concept by observing its "necessity to externalize the form of the Notion" (§806). This exploration is possible because we recognize that the limit of the Concept entails transcending beyond itself. Hegel elaborates:

> Self-knowing Spirit knows not only itself but also the negative of itself, or its limit: to know one's limit is to know how to sacrifice oneself. This sacrifice is the externalization in which Spirit displays the process of its becoming Spirit in the form of free contingent happening, intuiting its pure Self as Time outside of it, and equally its Being as Space.
>
> PoS §807, 492

Here, Hegel provides an explanation of the logical constitution of the concept as it is for-itself. It knows itself as a limit. Since the limit is always inherent to and transcendent of a specific object, it extends beyond itself. Consequently, the concept can apprehend itself as nature. Moving beyond the exploration of its internal necessity, which manifests as externalized nature, Hegel proceeds to elucidate the process of returning from the concept of nature to consciousness through spirit (§806). Collectively, these aspects form a circular framework: the self-knowing spirit, embodied as the concept, comprehends itself in an externalized form and can thereby grasp itself as nature. Spirit transforms into nature, an inherently existing substance or "living immediate Becoming," yet it surpasses its own existence. Through a series of sublations, it ultimately "reinstates the Subject," for without doing so, the fragmented nature could never be fully understood (PoS §807, 492).

Hegel's final stride entails establishing a connection between nature and history by means of the Concept of Spirit. History, according to Hoffmeyer (1992, 205), represents the self-externalization of Spirit. Additionally, as Hoffmeyer highlights, nature finds its existence only within the framework of history. This implies that the concept of nature can only be comprehended by recognizing its interdependence with historical subjects. Consequently, Hoffmeyer argues that nature is a concept that cannot persist solely as an immediate entity in itself; it must also undergo sublation. Hence, nature must transform into a subject. As Hegel later expounds in *The Philosophy of Spirit*, it is essential to delve into the anthropological constitution of humans, who manifest themselves within the context of history.

World Spirits exert influence over our self-conceptions. In relation to the preceding distinction between time as a sequence of spirits and the conceptual arrangement of these World Spirits, Hegel delineates two facets of the occurrence of the "goal, Absolute Knowing, or Spirit that knows itself as Spirit" along the "path of recollection of Spirits as they are in themselves" (PoS §808, 492). First, when considering these spirits from "the side of their free existence," they appear as "contingency" (PoS §808, 492). Hegel conceptualizes this temporal manifestation of Spirits as 'history' or, alternatively, 'world history'. This denotes the emergence of distinct time periods on a chronological timeline. Nevertheless, history is not merely a mere sequence of events but also encompasses the progressive development of concepts (see PoS §808, 492). Therefore, history emerges, as a second aspect from the comprehended organization of World Spirits (PoS §808, 492). Hegel refers to this as "the Science of Knowing in the sphere of appearance" (PoS §808, 492), also known as 'phenomenology'.

History and science taken together deliver "the actuality, truth, and certainty of his [Absolute Spirit's] throne" (PoS §808, 492). First, Spirit becomes actual, since the Absolute Knowing of Spirit is the culmination of its development toward self-knowledge. Second, Spirit is true, since Absolute Knowing offers a starting point, according to which subject and object are identical. Third, Spirit is certain, since Absolute Knowing's identity is based on the principle that guided the development of spirit. Moreover, Spirit is the form of conceptual movement that at its end knows itself as its limit concept, i.e., the ground on which everything is thought. However, by knowing its limit, it also goes beyond itself. Hegel expresses this through the idea of a sacrifice. This sacrifice, as the Concept externalizes itself, leads to nature. However, the negation of its negation—from Nature to Spirit—leads back to itself. It is the negating activity of the concept that brings it back to itself. During this process of negation, it comes to know itself, while the return to itself is a return that is higher than the starting point of the circle. We can also say that the circle never finishes its movement. This openness of the circular movement from Absolute Knowing, through Nature, to Spirit's history is one way of understanding the 'eternal' in Hegel.

Each new cycle must, however, emerge from a vantage point of greater elevation. Spirit, in this context, embodies the continuous flow of history that never comes to a halt. At the conclusion of each world-historical era, Spirit has the capacity to apprehend itself, yet this understanding necessitates the need to transcend itself. The process of conceptual thinking does not cease once a convincing thought is reached. When we examine our culture within the framework of history, we become aware of its flaws, mistakes, and injustices. Engaging in critical reflection on our culture implies that we cannot remain fixed at any given point, but rather, through reflection, our culture must change.

The resulting eternity of Spirit's movement is, therefore, limited to further experiences in time, which display themselves through consciousness. Since consciousness is part of this experiential circle, Absolute Knowing remains in history. It knows only how it constitutes an end of every single circle of world history. Hence, Absolute Knowing marks each new beginning; a moment in which we grasp our current World Spirit, present to us in its organization according to its own principle. In this moment, we grasp its opening toward a new history. In its openness, it occurs as an eternity. Yet, Absolute Knowing remains also as a goal. It is therefore the end of the former world history and the beginning of a new world history. In other words, grasping itself in its historical moment by means of reflection will remain the eternal task of reflecting subjects. It is the existential crisis of individuals to essentially yearn for the universal, yet so far find themselves confined within the boundaries of their own finiteness.

As we now know, there is no end to this existential reflection so far. Hence, instead of coming to a definite end of history, Hegel closes with Schiller's poetic lines: "from the chalice of this realm of spirits / foams forth for Him his own infinitude" (PoS §808, 493). With this metaphor, Hegel does not reduce science to a semantic explanation by virtue of the Concept. We cannot close the circle of experiences; we cannot close the project of knowledge with a last solution as Brandom, for example, postulates that we approach a reality in-itself. Historically, we can, however, know, if not who we will be, at least who we were and who we are for now. History is, thus, evolving with regard to the capacity for experience and reflection. These two ingredients –reflection and experience—build therefore the substance of subjective knowledge and spirit. Altogether, we can say that science has found its objective starting point, but science, which is still at its beginning, has not yet achieved its final form, and we do not know if it ever will.

CHAPTER 5

Conclusion on Brandom's Conceptual Realism and Hegel's Historicity

In contemporary discourse, there is a growing narrative suggesting that the longstanding philosophical divide may finally be on the verge of dissolution. Brandom's meticulous analytic examination of Hegel's *Phenomenology of Spirit* seems to lend credence to this notion. However, I contend that these recent claims of reconciliation often oversimplify the complexity of the ongoing debate.

Brandom, in particular, aims to tackle the challenges stemming from the early analytic tradition's problematic reading of Hegel. Initially, figures like Russell dismissed Hegel's purported anti-realism outright. Subsequently, in a bid to grapple with the issues surrounding discussions of mind-independent realism, Sellars and later Brandom repositioned Hegel as an inferentialist. However, their reintroduction of Hegel into the analytic discourse overlooks Hegelian historicism—a perspective that formulates a theory of knowledge within the confines of historical context, eschewing appeals to noumena or a reality independent of the mind.

My contention extends beyond the mere assertion of the incompatibility between Brandom's and Hegel's methodologies. I also advocate for a shift in the focus of the debate, emphasizing the significance of history over the concept of mind-independent reality. This assertion is rooted in the profound interconnection between knowledge and history. Consequently, I also challenge the notion of a rigid division between a philosophy of history and pure philosophy. Drawing from Hegelian principles, I posit that our claims to knowledge find their grounding only through a reflective exploration of history. To propel philosophy or any scientific discipline forward, we must build upon past theories, subjecting them to rigorous scrutiny and rendering them transparent. Hegel notably acknowledges that theories and their corresponding objects are products of this historical, dialectical development and extends it to the whole of our project of knowledge and society. Rather than positing a dichotomy between theories and mind-independent objects, we must perceive their intrinsic interconnectedness and explicate it through the lens of all our history.

Through this reflective journey, we attain an awareness of how theories shape our experiences and lay the groundwork for transcending their limitations. Hegel's concept of Absolute Knowing, representing the pinnacle of this

 | DOI:10.1163/9789004714809_006

reflection, entails the elucidation of all our theories and historical becoming. In reaching Absolute Knowing, we attain a comprehensive understanding of our history and can furnish explanations for every facet of our current historical experience.

It is crucial to acknowledge that Absolute Knowing, as encapsulated in the concept of The Concept, does not present a viewpoint removed from history, claiming to establish an ahistorical foundation conclusively. Rather, it signifies a historical grounding that offers a more effective orientation for the particular historical moment. Notably, Hegel's endeavor to comprehend all historical factors that shape us becomes increasingly elusive as societies evolve and history unfolds in ever-greater complexity. The integration of diverse cultures and the expansion of individual voices give rise to a multitude of historically consequential events. Indeed, this multitude of events and pertinent perspectives casts doubt on the feasibility of achieving such a comprehensive grounding for Absolute Knowing in our current historical epoch. Arguably, it is more challenging now than ever before.

Hegel's philosophy of Absolute Knowing may face challenges or appear insufficient in our present context, but such shortcomings do not stem from an inherent rigidity of Hegel's philosophy—often perceived as a single comprehensive, ahistoric framework called the system. In this regard, it is crucial to understand that Hegel does not posit the existence of a singular overarching super-spirit governing all sciences or dictating every historical event. So, for example, each scientific discipline operates within its own pragmatic framework, establishing positions and standards rooted in its unique historical development and objectives. Nonetheless, we can still shed light on the emergence and operation of these disciplines by considering them in the context of Absolute Spirit. Interpreting their evolution as a spiritual progression enables us to formulate initial, speculative insights that gesture towards a broader interconnectedness—Absolute Knowing. While achieving a comprehensive understanding of all factors shaping our existence in the present historical moment remains elusive, Hegel's overarching philosophy suggests that philosophy can persist as a discipline committed to providing a profound and unified reflection on the norms dictated by our historical context, oriented towards Absolute Knowing. Although it eschews reliance on external standards, Hegel's approach then constructs what Siep aptly terms 'slower-moving objects', which offer orientation within our complex historical contexts. From my perspective, Hegel's philosophy as merely oriented towards eternal Truth suggests a peculiar form of relativism captured in the concept of historicity. It embodies a relativism that provides and grounds practical, preliminary plateaus of knowledge with respect to the idea of the eternal, without being dogmatic or claiming ahistoricity. Most importantly, it acknowledges our innate yearning for infinite or eternal truth but also scrutinizes and contextualizes these yearnings within the confines of our finite historical perspective as human beings.[1] By this, it accomplishes an effective, relativist epistemological grounding—a theory of finite knowledge—that, nonetheless, remains open to the possibility of Truth. Never rejecting that eternal Truth remains a possibility, it is an open Spirit—open to eternal change, yet also open to recognizing Truth one day.

1 The Hegelian approach I highlight is notably absent in Baghramian's and Coliva's seminal work on relativism. While they do allocate a chapter to explore the historical underpinnings of relativism, their main exposition largely revolves around presenting relativism through an analytic vantage point. Notwithstanding this emphasis, the book compellingly illustrates how the arena of analytic philosophy has begun to welcome relativism in the recent decade, a concept that was once considered its archenemy (see Baghramian, Maria, and Annalisa Coliva. *Relativism*. London and New York: Routledge 2020).

Bibliography

Adorno, Theodor W. 1974. *Drei Studien zu Hegel.* Frankfurt: Suhrkamp.

Apel, Karl-Otto. 1988. *Diskurs und Verantwortung. Das Problem des Übergangs zur postkonventionellen Moral.* Frankfurt: Suhrkamp.

Austin, John L. 1972. *Zur Theorie der Sprechakte* [*How to do things with words*]. Translated by Eike von Savigny. Leipzig: Reclam.

Baghramian, Maria, and Annalisa Coliva. 2020. *Relativism.* London and New York: Routledge.

Baumann, Charlotte. 2019. "Hegel's Realm of Shadows: Logic as Metaphysics in the Science of Logic: by Robert Pippin." *British Journal for the History of Philosophy* 27 (6): 1256–1260. doi: 10.1080/09608788.2019.1616157.

Bavaresco, Agemir. 2012. "On Articulating Reasons of Robert Brandom and His Hegelian Methodology." *International Journal of Humanities and Social Science* 2 (19): 74–86. http://www.ijhssnet.com/journals/Vol_2_No_19_Special_Issue_October_2012/8.pdf.

Beiser, Frederick. 2005. *Hegel.* New York, London: Routledge.

Bell, Jeffrey A., Andrew Cutrofello and Paul M. Livingston. 2016. *Beyond the Analytic-Continental Divide: Pluralist Philosophy in the Twenty-First Century.* New York: Routledge.

Biletzki, Anat and Anat Matar. 2016. "Ludwig Wittgenstein." *The Stanford Encyclopedia of Philosophy*. Edited by Edward N. Zalta. Accessed November 5, 2017. http://plato.stanford.edu/archives/fall2016/entries/wittgenstein.

Bohman, James and William Rehg. 2017. "Jürgen Habermas." *The Stanford Encyclopedia of Philosophy.* Edited by Edward N. Zalta. Accessed on November 26, 2017 https://plato.stanford.edu/archives/fall2017/entries/habermas/.

Brandom, Robert. 1997a. "Review: Précis of Making It Explicit." *Philosophy and Phenomenological Research* 57 (1): 153–156. http://www.jstor.org/stable/2953784.

Brandom, Robert. 1997b. "Replies." *Philosophy and Phenomenological Research* 57 (1): 189–204. doi:10.2307/2953789.

Brandom, Robert. 1998. *Making It Explicit.* Cambridge: Harvard University Press.

Brandom, Robert. 2000a. *Articulating Reasons: An Introduction to Inferentialism.* Cambridge: Harvard University Press.

Brandom, Robert. 2000b. "Vocabularies of Pragmatism: Synthesizing Naturalism and Historicism." *Rorty and His Critics.* Edited by Robert Brandom, 156–82. Oxford: Blackwell Publishers.

Brandom, Robert. 2002. *Tales of the Mighty Dead: Historical Essays in the Metaphysics of Intentionality.* Cambridge: Harvard University Press.

Brandom, Robert. 2007. "Inferentialism and Some of Its Challenges." *Philosophy and Phenomenological Research* LXXIV (3): 651–676. doi: 10.1111/j.1933-1592.2007.00044.x.

Brandom, Robert. 2008a. *Between Saying and Doing: Towards an Analytic Pragmatism*. Oxford: Oxford University Press.

Brandom, Robert. 2008b. "Responses." *The Pragmatics of Making it Explicit.* Edited by Pirmin Stekeler-Weithofer. Amsterdam: John Benjamins Publishing.

Brandom, Robert. 2009. *Reason in Philosophy: Animating Ideas*. Cambridge: Harvard University Press.

Brandom, Robert. 2010. "Reply to Charles Taylor's 'Language not mysterious?'" *Reading Brandom. On Making It Explicit.* Edited by Bernhard Weiss and Jeremy Wanderer. New York: Routledge.

Brandom, Robert. 2013. "A Sketch of a Program for a Critical Reading of Hegel: Comparing Empirical and Logical Concepts." *Recent Lecture Series and Collections*. Accessed on September 9, 2024. https://sites.pitt.edu/~rbrandom/Texts%20Mark%201%20p.html.

Brandom, Robert. 2019. *A Spirit of Trust: A Reading of Hegel's Phenomenology*. Cambridge: Harvard University Press.

Bransen, Jan. 2002. "Normativity as the Key to Objectivity—An exploration of Robert Brandom's Articulating Reasons." *Inquiry* 45 (3): 373–392. https://doi.org/10.1080/002017402760258204.

Braun, Edmund. 1996. *Der Paradigmenwechsel in der Sprachphilosophie*. Darmstadt: Wissenschaftliche Buchgesellschaft Darmstadt.

Braun, Edmund. 2007. "K.-O. Apel—Transzendentalpragmatik als normativ-semiotische Transformation der Transzendentalphilosophie." *Aufgang—Jahrbuch für Denken, Dichten, Musik* 4. Edited by Jose Sanches de Murillo and Martin Thurner. Stuttgart: W. Kohlhammer.

Chase, James, and Jack Reynolds. 2014. *Analytic Versus Continental: Arguments on the Methods and Value of Philosophy*. New York: Routledge.

Chung, I-Lin. 2008. "No Communal and First-Person Errors: A Critique of Brandom's Objectivity Proof." *National Cheng Chi University Philosophical Journal* 20: 105–122. Accessed September 19, 2024. https://thinker.nccu.edu.tw/upload/40/publication_file/551/林從一(Chung-I%20Lin)20.pdf.

DeVries, Willem A. 2005. *Wilfrid Sellars*. Chesham: Acumen Publishing.

DeVries, Willem A. 2009. "Getting Beyond Idealisms." *Empiricism, Perceptual Knowledge, Normativity, and Realism*. Edited by Willem A. DeVries. Oxford: Oxford University Press.

DeVries, Willem A. 2016. "Wilfrid Sellars." *The Stanford Encyclopedia of Philosophy*. Edited by Edward N. Zalta. https://plato.stanford.edu/archives/win2016/entries/sellars/.

DeVries, Willem A. 2017. "Hegel's Revival in Analytic Philosophy." *The Oxford Handbook of Hegel.* Edited by Dean Moyar. doi: 10.1093/oxfordhb/9780199355228.013.35.

DeVries, Willem A. 2021a. "Hegel Today." *Aeon*. Edited by Sam Dresser. https://aeon.co/essays/how-the-anglophone-world-is-rediscovering-hegels-philosophy.

DeVries, Willem A. 2021b. "Brandom and *A Spirit of Trust*." *International Journal of Philosophical Studies* 29 (2): 236–250. doi: 10.1080/09672559.2021.1918822.

Diggins, John Patrick. 1991. *The Promise of Pragmatism*. Chicago: University of Chicago Press.

Düsing, Klaus. 1976. *Das Problem der Subjektivität in Hegels Logik*. Bonn: Bovier Verlag, Herbert Grundmann.

Erp, H.V. 1998. "The End of History and Hegel's Conception of Modernity." *Ideas Y Valores* 47 (107) http://www.revistas.unal.edu.co/index.php/idval/article/view/21868.

Ferraris, Maurizio. 2014. "Introduction." *Bridging the Analytical Continental Divide: A Companion to Contemporary Western Philosophy*. Edited by Tiziana Andina. Leiden: Brill.

Field, Hartry. 2009. "Epistemology without Metaphysics." *Philosophical Studies* 143 (2): 249–290. doi:10.1007/s11098-009-9338-1.

Fierke, K.M. 2015. "Breaking the Silence." *Language, Agency, and Politics in a Constructed World*. Edited by Francois Debrix. Oxon: Routledge.

Fodor, Jerry and Ernie Lepore. 2001 "Brandom's Burdens: Compositionality and Inferentialism." *Philosophy and Phenomenological Research* 63 (2): 465–481. doi:10.1111/j.1933-1592.2001.tb00119.x.

Fodor, Jerry and Ernie Lepore. 2007. "Brandom Beleaguered." *Philosophy and Phenomenological Research* LXXIV (3): 677–691. https://www.jstor.org/stable/40041074.

Franks, Paul W. 2005. *All or Nothing: Systematicity, Transcendental Arguments, and Skepticism in German Idealism*. Cambridge: Harvard University Press.

Fulda, Hans Friedrich. 2007. "Das absolute Wissen—sein Begriff, Erscheinen und Wirklichwerden." *Revue de Métaphysique et de Morale* 55 (3): 338–401. doi: 10.3917/rmm.073.0338.

Gadamer, Hans-Georg. 1976. *Hegel's Dialectic: Five Hermeneutical Studies*. Translated by P. Christopher Smith. New Haven: Yale University Press.

Gansel, Cristina, and Frank Jürgens. 2002. *Textlinguistik und Textgrammatik*. Wiesbaden: Westdeutscher Verlag.

Gee, James Paul. 1999. *An Introduction to Discourse Analysis—Theory and Method*. London: Routledge. 1999.

Gerhard, Myriam. 2015. *Hegel und die logische Frage*. Berlin: Walter de Gruyter.

Giovagnoli, Raffaela. 2001. "On Normative Pragmatics: A Comparison between Brandom and Habermas." *Teorema* 20 (3): 51–68. https://www.jstor.org/stable/43046449.

Gleeson, Jules. 2020. "Robert Brandom, a Philosopher's Philosopher." *JstorDaily*. 2020. Accessed June 13, 2021, https://daily.jstor.org/robert-brandom-a-philosophers-philosopher/.

Greene, Murray. 1972. *Hegel on the Soul—A speculative ontology*. Hague: Martinus Nijhoff.

Grier, Phillip T. 2017. *Identity and Difference: Studies in Hegel's Logic, Philosophy of Spirit, and Politics*. New York: Suny Press.

Grönert, Peter. 2005. "Brandom's Solution to the Objectivity Problem." *Pragmatics & Cognition* 13 (1): 161–175. doi: https://doi.org/10.1075/pc.13.1.12gro.

Habermas, Jürgen. 1976. *Zur Rekonstruktion des Historischen Materialismus*. Frankfurt: Suhrkamp.

Habermas, Jürgen. 1987. *Knowledge and Human Interest*. Translated by Jeremy J. Shapiro. New York: Polity Press.

Habermas, Jürgen. 1991. *Erläuterungen zur Diskursethik*. Frankfurt: Suhrkamp.

Habermas, Jürgen. 1999. *Wahrheit und Rechtfertigung*. Frankfurt: Suhrkamp.

Habermas, Jürgen. 2003. *Truth and Justification*. Translated by Barbara Fultner. Cambridge, Massachusetts: MIT Press.

Habermas, Jürgen. 2005. *Die Zukunft der menschlichen Natur: Auf dem Weg zu einer liberalen Eugenik?* Frankfurt: Suhrkamp.

Habermas, Jürgen. 2015. *On the Logic of the Social Sciences*. Translated by Shierry Weber Nicholson and Jerry Stark. Cambridge: Polity Press.

Hale, Bob and Crispin Wright. 2010. "Assertibilist Truth and Objective Content—Still Inexplicit?" *Reading Brandom. On Making It Explicit*. Edited by Bernhard Weiss and Jeremy Wanderer. New York: Routledge.

Harris, H. S. 1997. *Hegel's Ladder. Volume II: The Odyssey of Spirit*. Indianapolis: Hackett Publishing.

Hegel, G.W.F. 1861. *Lectures on the Philosophy of History*. Translated by J. Sibree. London: Henry G. Bohn.

Hegel, G.W.F. 1954. *Philosophische Bibliothek 237*. Hamburg: Meiner.

Hegel, G.W.F. 1959. *Philosophy of Nature: Encyclopedia of the Philosophical Sciences II (1830)*. Translated by A.V. Miller. Oxford: Oxford University Press.

Hegel, G.W.F. 1969. *Wissenschaft der Logik I, II*. Frankfurt: Suhrkamp.

Hegel, G.W.F. 1977a. *Phenomenology of Spirit*. Translated by A.V. Miller. Oxford: Oxford University Press.

Hegel, G.W.F. 1977b. *The Difference Between Fichte's and Schelling's System of Philosophy: An English Translation of G. W. F. Hegel's Differenz des Fichte'schen und Schelling'schen Systems der Philosophie*. Edited and Translated by Walter Cerf and H.S. Harris. Albany: State of New York Press.

Hegel, G.W.F. 1986. "Vorlesungen über die Geschichte der Philosophie I." *Gesammelte Werke 18*. Edited by Eva Moldenhauer und Karl Markus Michel. Frankfurt: Suhrkamp.

Hegel, G.W.F. 1993. *Vorlesungen über die Geschichte der Philosophie—Einleitung, Orientalische Philosophie.* Edited by Walter Jaeschke. Hamburg: Meiner.

Hegel, G.W.F. 1995. *Lectures on the History of Philosophy.* Translated by E.S. Haldane. Lincoln: Nebraska Press.

Hegel, G.W.F. 2010. *Science of Logic.* Translated by George Di Giovanni. Cambridge: Cambridge University Press.

Hegel, G.W.F. 2014a. *Enzyklopädie der philosophischen Wissenschaften im Grundrisse 1830; Erster Teil—Die Wissenschaft der Logik mit den Mündlichen Zusätzen.* Frankfurt: Suhrkamp.

Hegel, G.W.F. 2014b. *Enzyklopädie der philosophischen Wissenschaften im Grundrisse 1830; Zweiter Teil—Naturphilosophie.* Frankfurt: Suhrkamp.

Hegel, G.W.F. 2015. "Vorlesungen über die Philosophie der Weltgeschichte. Band 1: Die Vernunft in der Geschichte." *Gesammelte Werke* 27. Edited by Bernadette Collenberg-Plotnikov. Hamburg: Meiner.

Hempel, Carl Gustav. 1950. "Problems and Changes in the Empiricist Criterion of Meaning." *Revue Internationale de Philosophie* 4 (11) (1950): 41–63. http://www.jstor.org/stable/23932368.

Höffe, Ottfried. 2010. *Kant's Critique of Pure Reason: The Foundation of Modern Philosophy.* Tübingen: Springer.

Hoffmeyer, John F. 1992. "Absolute Knowing and the Historicity of Spirit." *The Journal of Religion* 72 (2): 198–209. http://www.jstor.org/stable/1205149.

Hookway, C. J. 2016. "Pragmatism." *The Stanford Encyclopedia of Philosophy.* Edited by Edward N. Zalta. Accessed November 25, 2017. https://plato.stanford.edu/archives/sum2016/entries/pragmatism.

Houlgate, Stephen. 2020. "Brandom, *A Spirit of Trust: A Reading of Hegel's Phenomenology*" *Notre Dame Philosophical Reviews.* Accessed June 13, 2021. https://ndpr.nd.edu/reviews/a-spirit-of-trust-a-reading-of-hegels-phenomenology/.

Hoyningen-Huene, Paul. 2013. *Systematicity: The Nature of Science.* Oxford: Oxford University Press.

Huetter-Almerigi, Yvonne. 2020. "Two Forms of Realism—Making Sense of Rorty's Controversy with Brandom and Ramberg over Objectivity." *European Journal of Pragmatism and American Philosophy* [Online], XII-1:1–14, https://journals.openedition.org/ejpap/1868.

Irvine, Andrew David. 2015. "Bertrand Russell." *The Stanford Encyclopedia of Philosophy.* Edited by Edward N. Zalta. Accessed July 5, 2017. https://plato.stanford.edu/archives/win2015/entries/russell.

Jackman, Henry. 2017. "Meaning Holism." *The Stanford Encyclopedia of Philosophy.* Edited by Edward N. Zalta. Accessed on November 25, 2017. https://plato.stanford.edu/archives/spr2017/entries/meaning-holism.

Jaeschke, Walter. 1984. "World History and the History of Absolute Spirit." *History and System: Hegel's Philosophy of History*. Edited by Robert L. Perkins. Albany: State of New York Press.

Jaeschke, Walter. 1993. "Einleitung." *Vorlesungen über die Geschichte der Philosophie—Einleitung, Orientalische Philosophie by G.W.F. Hegel.* Edited by Walter Jaeschke. Hamburg: Meiner.

Jaeschke, Walter. 2020. *Hegels Philosophie.* Hamburg: Meiner.

Kant, Immanuel. 1990. *Kritik der Reinen Vernunft.* Hamburg: Felix Meiner.

Kant, Immanuel. 1998. *Critique of Pure Reason*. Translated by Paul Guyer. Cambridge: Cambridge University Press.

Kolb, David. 1978. "Sellars and the Measure of All Things." *Philosophical Studies: An International Journal for Philosophy in the Analytic Tradition* 34 (4): 381–400. http://www.jstor.org/stable/4319263.

Kreines, James, and Rachel Zuckert. 2017. *Hegel on Philosophy in History.* Cambridge: Cambridge University Press.

Kremer, Michael. 2010. "Representation or Inference. Must We Choose? Should We?" *Reading Brandom. On Making It Explicit.* Edited by Bernhard Weiss and Jeremy Wanderer. New York: Routledge.

Kuhlmann, Wolfgang. 1992. *Sprachphilosophie—Hermeneutik—Ethik.* Würzburg: Königshausen & Neumann.

Kusch, Martin. 2002. *Knowledge by Agreement.* Oxford: Clarendon Press.

Lafont, Christina. 1999. *The Linguistic Turn in Hermeneutic Philosophy.* Translated by Jose Medina. Cambridge: MIT Press.

Lakoff, George, and Mark Johnson. 1980. *Metaphors we live by.* Chicago: University of Chicago Press.

Levine, Steven. 2007. "The Place of Picturing in Sellars' Synoptic Vision." *The Philosophical Forum 38* (3): 247–269. doi:10.1111/j.1467-9191.2007.00266.

Levine, Steven. 2010. "Habermas, Kantian Pragmatism, and Truth." *Philosophy and Social Criticism* 36 (6): 677–695. http://journals.sagepub.com/doi/abs/10.1177/0191453710366216.

Levine, Steven. 2011. "Truth and Moral Validity: On Habermas' Domesticated Pragmatism." *Constellations* 18 (2): 244–259. doi/10.1111/j.1467-8675.2011.00636.x.

Levine, Steven. 2012. "Brandom's Pragmatism." *Transactions of the Charles S. Peirce Society* 48 (2): 125–140. https://doi.org/10.2979/trancharpeirsoc.48.2.125.

Long, Joseph W. 2002. "Who's a Pragmatist: Distinguishing Epistemic Pragmatism and Contextualism." *The Journal of Speculative Philosophy* 16 (1): 39–49. http://www.jstor.org/stable/25670401.

Mander, W. J. 2011. *British Idealism: A History*. Oxford: Oxford University Press.

Margolis, Joseph. 2007. *Pragmatism without Foundations: Reconciling Realism and Relativism*. London: Continuum.

Margolis, Joseph. 2009. "A 'Pragmatist' among Disputed Pragmatists: Robert Brandom's Between Saying and Doing: Towards an Analytic Pragmatism." *Contemporary Pragmatism* 6 (1): 185–196.

Marshall, David L. 2013. "The Implications of Robert Brandom's Inferentialism for Intellectual History." *History and Theory* 52: 1–31. doi: https://doi.org/10.1111/hith.10651.

Marx, Werner. 1975. *Hegel's Phenomenology of Spirit*. Translated by Peter Heath. Chicago: University of Chicago Press.

McDowell, John. 1996. *Mind and World*. Cambridge: Harvard University Press.

McDowell, John. 1997. "Brandom on Representation and Inference." *Philosophy and Phenomenological Research* 57 (1): 157–162.

McDowell, John. 2017. "Why Does it Matter to Hegel that Geist has a History?" *Hegel on Philosophy in History*. Edited by James Kreines and Rachel Zuckert. Cambridge: Cambridge University Press.

Mosteller, Timothy M. 2014. *Theories of Truth: An Introduction*. London: Bloomsbury Publishing.

Mueller, Alex. 2014. "Löst Brandoms Inferentialismus bedeutungsholistische Kommunikationsprobleme?" *Zeitschrift für Semiotik* 34 (3–4): 141–185. Accessed July 02, 2017. https://philpapers.org/archive/MUELBI.pdf.

Murzi, Julien, and Florian Steinberger. 2017. "Inferentialism." *A Companion to the Philosophy of Language*. Edited by Bob Hale, Crispin Wright and Alexander Miller. Malden, Oxford: Wiley Blackwell. https://doi.org/10.1002/9781118972090.ch9

Nickel, Bernhard. 2011. "Dynamics, Brandom-Style." *Philosophical Studies* 162 (2): 333–354. Accessed May 29, 2017. https://dash.harvard.edu/handle/1/13041028.

Papineau, David. 2017. "Naturalism." *The Stanford Encyclopedia of Philosophy*. Edited by Edward N. Zalta. Accessed May 1st, 2017. http://plato.stanford.edu/archives/fall2015/entries/naturalism/.

Pedersen, Jørgen. 2008. "Habermas' Method: Rational Reconstruction." *Philosophy of the Social Sciences* 38 (4): 457–485. http://journals.sagepub.com/doi/abs/10.1177/0048393108319024.

Peirce, Charles Sanders. 1998. *The Essential Peirce, Volume 2: Selected Philosophical Writings (1893–1913)*. Bloomington: Indiana University Press.

Pihlström, Sami. 2015. "Pragmatist Metaphysics." *The Bloomsbury Companion to Pragmatism*. Edited by Sami Philström. London: Bloomsbury Publishing.

Pinkard, Terry. 1994. *Hegel's Phenomenology—The Sociality of Reason*. Cambridge: Cambridge University Press.

Pinkard, Terry. 2014. "Semantic Self-Consciousness: Brandom On Hegel (paper given at a conference: Language and Modernity: Brandom's Semantic Reading of Hegel's Phenomenology of Spirit). (Draft)." Accessed on June 14, 2021. https://www.academia.edu/7917240/Semantic_Self_Consciousness_Brandom_On_Hegel.

Pinkard, Terry. 2020. "A Spirit of Trust: A Reading of Hegel's Phenomenology, by Robert Brandom." *Mind* 129 (515): 990–999. https://doi.org/10.1093/mind/fzz060, (https://www.academia.edu/40679479/Book_review_A_Spirit_of_Trust_A_Reading_of_Hegels_Phenomenology_by_Robert).

Pippin, Robert. 1989. *Hegel's Idealism. The Satisfactions of Self-Consciousness*. Cambridge: Cambridge University Press.

Pippin, Robert. 1990. "Hegel and Category Theory." *The Review of Metaphysics*. 43 (4): 839–848. https://www.jstor.org/stable/20128948.

Pippin, Robert. 1997. *Idealism as Modernism*. Cambridge: Cambridge University Press.

Pippin, Robert. 2005. "Brandom's Hegel." *European Journal of Philosophy* 13 (3): 381–408. https://doi.org/10.1111/j.1468-0378.2005.00235.x.

Pippin, Robert. 2008. *Hegel's Practical Philosophy: Rational Agency as Ethical Life*. Cambridge: University Press.

Prien, Bernd. 2010. "Robert Brandom on Communication, Reference, and Objectivity." *International Journal of Philosophical Studies* 18 (3): 433–458. http://dx.doi.org/10.1080/09672559.2010.494439.

Priest, Graham and Francesco Berto. 2017. "Dialetheism." *The Stanford Encyclopedia of Philosophy*. Edited by Edward N. Zalta. Accessed November 26, 2017. https://plato.stanford.edu/archives/spr2017/entries/dialetheism/.

Putnam, Hilary. 1998. "A Half Century of Philosophy Viewed from Within." *American Academic Culture in Transformation: Fifty Years, Four Disciplines*. Edited by Thomas Bender and Carl Schorske, 193–227. Princeton: Princeton University Press.

Quante, Michael. 2001. "Vorwort." *Vermittlung und Versöhnung: die Aktualität von Hegels Denken für ein zusammenwachsendes Europa*. Edited by Michael Quante and Erzsébet Rózsa. Münster: Lit Verlag Münster.

Quine, Willard Van Orman. 1951. "Main Trends in Recent Philosophy: Two Dogmas of Empiricism." *The Philosophical Review* 60 (1): 20–43. http://www.jstor.org/stable/2181906.

Radford, Luis. 2017. "On Inferentialism." *Mathematics Education Research Journal* 29: 493–508. https://doi.org/10.1007/s13394-017-0225-3.

Redding, Paul. 2007. *Analytic Philosophy and the Return of Hegelian Thought*. Cambridge: Cambridge University Press.

Reinikainen, Jaako. 2021. "Brandom and the Pragmatic Quest for Semantic Objectivity." *Acta Philosophica Fennica* 97: 55–78, 2021. https://philarchive.org/archive/REIBAT-4v1.

Rescher, Nicholas. 1979. *Cognitive Systematization: A Systems-theoretic Approach to a Coherentist Theory of Knowledge*. Totowa, NJ: Rowman and Littlefield.

Rescher, Nicholas. 2005. *Cognitive Harmony: The Role of Systemic Harmony in the Constitution of Knowledge*. Pittsburgh: Pittsburgh University Press.

Rockmore, Tom. 1989. *Habermas on Historical Materialism.* Bloomington: Indiana University Press.

Rockmore, Tom. 1993. "Some Problems in Recent Pragmatism." *History of Philosophy Quarterly* 10 (3): 277–92. http://www.jstor.org/stable/27744065.

Rockmore, Tom. 1996. *On Hegel's Epistemology and Contemporary Philosophy*. New Haven: Humanity Press.

Rockmore, Tom. 1997. *Cognition*. Berkeley: University of California Press.

Rockmore, Tom. 2000. "Essentialism, Phenomenology, and Historical Cognition." *The Empirical and the Transcendental: A Fusion of Horizons*. Edited by Bina Gupta. Oxford: Rowman and Littlefield Publishers.

Rockmore, Tom. 2001. "Analytic Philosophy and the Hegelian Turn." *The Review of Metaphysics* 55 (2): 339–370. http://www.jstor.org/stable/20131711.

Rockmore, Tom. 2004a. *Foundationalism*. Oxford: Rowman and Littlefield Publishers.

Rockmore, Tom. 2004b. "On the Structure of Twentieth-Century Philosophy." *Metaphilosophy* 35 (4): 466–478. http://www.jstor.org/stable/24439712.

Rockmore, Tom. 2005. *Hegel, Idealism, and Analytic Philosophy.* New Haven: Yale University Press.

Rockmore, Tom. 2007. *Kant and Idealism*. New Haven: Yale University Press.

Rockmore, Tom. 2008. *In Kant's Wake: Philosophy in the Twentieth Century*. Oxford: Blackwell Publishing.

Rockmore, Tom, 2010. *Kant and Phenomenology.* London and Chicago: University of Chicago Press.

Rockmore, Tom. 2012. "The Pittsburgh School, The Given and Knowledge." *Social Epistemology Review and Reply Collective* 2 (1): 29–38. Accessed November 11, 2017. http://wp.me/p1Bfg0-Ak.

Rockmore, Tom. 2016. "On the Pittsburgh School, Kant, Hegel, and Realism." *Wilfrid Sellars, Idealism, and Realism: Understanding Psychological Nominalism*. Edited by Patrick J. Reider. London: Bloomsbury Publishing.

Rodriguez-Pereyra, Gonzalo. 2015. "Nominalism in Metaphysics." *The Stanford Encyclopedia of Philosophy*. Edited by Edward N. Zalta. Accessed November 22, 2017. http://plato.stanford.edu/archives/sum2015/entries/nominalism-metaphysics.

Rorty, Richard. 1982a. "Philosophy in America Today." *The American Scholar* 51(2): 183–220.

Rorty, Richard. 1982b. *Consequences of Pragmatism.* Minneapolis: University of Minnesota Press.

Rorty, Richard. 1988. "Representation, Social Practise, and Truth." *Philosophical Studies: An International Journal for Philosophy in the Analytic Tradition* 54 (2): 215–228. http://www.jstor.org/stable/4319982.

Rorty, Richard. 1997a. "'Introduction' to Wilfrid Sellars." *Empiricism and the Philosophy of Mind.* Edited by Richard Rorty. Cambridge: Harvard University Press.

Rorty, Richard. 1997b. "What Do You Do When They Call You a 'Relativist'?" *Philosophy and Phenomenological Research* 57 (1): 173–177. http://www.jstor.org/stable/2953787.

Rorty, Richard. 1998. *Truth and Progress—Philosophical Papers.* Cambridge: Cambridge University Press.

Rorty, Richard. 2000. "Response to Robert Brandom." *Rorty and His Critics*. Edited by Robert Brandom. Oxford: Blackwell Publishers.

Rorty, Richard. 2003. "Analytic and Conversational Philosophy" *A House Divided—Comparing Analytic and Continental Philosophy*. Edited by C.G. Prado. Amherst, NJ: Humanity Books.

Rorty, Richard. 2021. *Pragmatism as Anti-Authoritarianism.* Cambridge: Harvard University Press.

Rosenkranz, Sven. 2001. "Farewell to Objectivity: A Critique of Brandom." *The Philosophical Quarterly* 51 (203): 232–237. doi:10.1111/j.0031-8094.2001.00226.x.

Russell, Bertrand. 1914. *Our Knowledge of the External World.* Chicago: Open Court.

Russell, Bertrand. 1995. *My Philosophical Development.* London: Routledge.

Rydenfelt, Henrik. 2019. "Realism without Representationalism." *Synthese* 198: 2901–2918, https://doi.org/10.1007/s11229-019-02251-4.

Sachs, Carl B. 2020. "Critical notice of Robert Brandom, A Spirit of Trust. A Reading of Hegel's Phenomenology." *European Journal of Pragmatism and American Philosophy* XII (2): 1–7. https://doi.org/10.4000/ejpap.2231.

Schalhorn, Christof. 2000. *Hegels enzyklopädischer Begriff von Selbstbewußtsein.* Hamburg: Meiner.

Schmied-Kowarzik, Wolfdietrich. 1999. *Denken aus geschichtlicher Verantwortung: Wegbahnungen zur praktischen Philosophie.* Würzburg: Königshausen & Neumann.

Sedgwick, Sally. 2017. "Remarks on History, Contingency, and Necessity in Hegel's Logic" *Hegel on Philosophy in History*. Edited by James Kreines and Rachel Zuckert. Cambridge: Cambridge University Press.

Sellars, Wilfrid. 1949. "Language, Rules and Behavior." *John Dewey: Philosopher of Science and Freedom*. Edited by Sidney Hook. New York: The Dial Press. http://www.ditext.com/sellars/lrb.html.

Sellars, Wilfrid. 1956. "Is There a Synthetic Apriori?" *American Philosophers at Work: The Philosophic Scene in the United States*. Edited by Sidney Hook. New York: Criterion Books.

Sellars, Wilfrid. 1962. "Truth and 'Correspondence'." *The Journal of Philosophy* 59 (2): 29–56. doi:10.2307/2023294.

Sellars, Wilfrid. 1968. *Science and Metaphysics—Variations on Kantian Themes.* New York: Humanities Press.

Sellars, Wilfrid. 1974. "Some Remarks on Kant's Theory of Experience." *Essays in Philosophy and its History.* Dordrecht-Holland: D. Reidel Publishing Company.

Sellars, Wilfrid. 1975. "Autobiographical Reflections." *Action, Knowledge, and Reality: Critical Studies in Honor of Wilfrid Sellars.* Edited by Hector-Neri Castañeda. Indianapolis: The Bobbs-Merrill Company. http://www.ditext.com/sellars/ar.html.

Sellars, Wilfrid. 1997. *Empiricism and a Philosophy of Mind.* Cambridge: Harvard University Press.

Sellars, Wilfrid. 2007. "Being and Being Known." *The Space of Reasons, Selected Essays of Wilfrid Sellars.* Edited by Kevin Scharp and Robert Brandom. Cambridge: Harvard University Press.

Sicha, Jeffrey F. 2014. "Defending the Unpopular Sellars: Picturing and 'The Descriptive'." *Journal of Philosophical Research* 39: 127–163. doi 10.5840/jpr201472914.

Siep, Ludwig. 2017. "Philosophy and the 'Stream' of Cultural History" *Hegel on Philosophy in History.* Edited by James Kreines and Rachel Zuckert. Cambridge: Cambridge University Press.

Stout, Jeffrey. 2004. *Democracy and Tradition.* Oxford: Princeton University Press.

Stekeler-Weithofer, Pirmin. 2008. *The Pragmatics of Making it Explicit.* Amsterdam: John Benjamins Publishing.

Strassfeld, Jonathan. 2020. "American Divide: The Making of 'Continental' Philosophy" *Modern Intellectual History* 17 (3): 833–866.

Swindal, James. 2012. *Action and Existence—A Case for Agent Causation.* New York: Palgrave Macmillan.

Taylor, Charles. 2010. "Language not Mysterious?" *Reading Brandom—On Making It Explicit.* Edited by Bernhard Weiss and Jeremy Wanderer, 32–47. New York: Routledge.

Veluwenkamp, Herman. 2021. "Inferentialist Truth Pluralism." *Ethical Theory and Moral Practice* 24: 107–120.

Weiss, Bernhard and Jeremy Wanderer. 2010. *Reading Brandom—On Making It Explicit.* New York: Routledge.

Westphal, Merold. 1979. *History and Truth in Hegel's Phenomenology.* Bloomington: Indiana University Press.

White, Alan. 1983. *Absolute Knowledge: Hegel and the Problem of Metaphysics.* Athens: Ohio University Press.

Whiting, Daniel. 2008. "Meaning Holism and De Re Ascription." *Canadian Journal of Philosophy* 38: 575–599. http://www.tandfonline.com/doi/abs/10.1353/cjp.0.0033.

Williams, Jeffrey J. 2013. "Inferential Man: An Interview with Robert Brandom." *Symploke* 21 (1–2): 367–391. doi:10.5250/symploke.21.1-2.0367.

Williams, Michael. 2016. "Pragmatism, Sellars and Truth." *Sellars and his Legacy.* Edited by James R. O'Shea, 223–260. Oxford: Oxford University Press.

Williamson, Timothy. 2010. "Review of Robert Brandom, *Reason in Philosophy: Animating Ideas.*" *Times Literary Supplement* 5579: 22–23, http://media.philosophy.ox.ac.uk/assets/pdf_file/0004/35833/tlsbrandom.pdf.

Wittgenstein, Ludwig. 1922. *Tractatus Logico-Philosophicus.* New York: Harcourt, Brace and Company.

Wolff, Jonathan. 2017. "Karl Marx." *The Stanford Encyclopedia of Philosophy.* Edited by Edward N. Zalta. Accessed November 26, 2017, https://plato.stanford.edu/archives/win2017/entries/marx.

Žižek, Slavoj. 2020. "Hegel On The Future, Hegel In The Future." *Philosophy Now* 140. https://philosophynow.org/issues/140/Hegel_On_The_Future_Hegel_In_The_Future.

Index